Mathematics Manual for Water and Wastewater Treatment Plant Operators

**Second Edition**

# Water Treatment Operations

## Math Concepts and Calculations

# Mathematics Manual for Water and Wastewater Treatment Plant Operators, Second Edition

---

*Basic Mathematics for Water and Wastewater Operators*

*Water Treatment Operations: Math Concepts and Calculations*

*Wastewater Treatment Operations: Math Concepts and Calculations*

Mathematics Manual for Water and Wastewater Treatment Plant Operators

**Second Edition**

# Water Treatment Operations

## Math Concepts and Calculations

**Frank R. Spellman**

CRC Press
Taylor & Francis Group
Boca Raton London New York

CRC Press is an imprint of the
Taylor & Francis Group, an **informa** business

CRC Press
Taylor & Francis Group
6000 Broken Sound Parkway NW, Suite 300
Boca Raton, FL 33487-2742



Printed on acid-free paper
Version Date: 20140114



International Standard Book Number-13: 978-1-4822-2421-4 (Paperback)



**Library of Congress Cataloging-in-Publication Data**

Spellman, Frank R.
Mathematics manual for water and wastewater treatment plant operators / author, Frank R. Spellman. -- Second edition.
volumes cm
Includes bibliographical references and index.
Contents: volume 1. Basic mathematics for water and wastewater operators -- volume 2. Water treatment operations : math concepts and calculations -- volume 3. wastewater treatment operations : math concepts and calculations.
ISBN 978-1-4822-2421-4 (paperback)
1. Water--Purification--Mathematics. 2. Water quality management--Mathematics. 3. Water--Purification--Problems, exercises, etc. 4. Water quality management--Problems, exercises, etc. 5. Sewage--Purification--Mathematics. 6. Sewage disposal--Mathematics. 7. Sewage--Purification--Problems, exercises, etc. 8. Sewage disposal--Problems, exercises, etc. I. Title.

TD430.S64 2014
628.101'51--dc23 2013050298

**Visit the Taylor & Francis Web site at**
**http://www.taylorandfrancis.com**

**and the CRC Press Web site at**
**http://www.crcpress.com**

# Contents

# Preface

Hailed on its first publication as a masterly account written in an engaging, highly readable, user-friendly style, the *Mathematics Manual for Water and Wastewater Treatment Plant Operators* has been expanded and divided into three specialized texts that contain hundreds of worked examples presented in a step-by-step format; they are ideal for all levels of water treatment operators in training and practitioners studying for advanced licensure. In addition, they provide a handy desk reference and handheld guide for daily use in making operational math computations. *Basic Mathematics for Water and Wastewater Operators* covers basic math operators and operations, *Water Treatment Operations: Math Concepts and Calculations* covers computations used in water treatment, and *Wastewater Treatment Operations: Math Concepts and Calculations* covers computations commonly used in wastewater treatment plant operations.

To properly operate a waterworks or wastewater treatment plant and to pass the examination for a waterworks/wastewater operator's license, it is necessary to know how to perform certain calculations. In reality, most of the calculations that operators at the lower level of licensure need to know how to perform are not difficult, but all operators must have a basic understanding of arithmetic and problem-solving techniques to be able to solve the problems they typically encounter.

How about waterworks/wastewater treatment plant operators at the higher levels of licensure? Do they also need to be well versed in mathematical operations? The short answer is absolutely. The long answer is that those working in water or wastewater treatment who expect to have a successful career that includes advancement to the highest levels of licensure or certification (usually prerequisites for advancement to higher management levels) must have knowledge of math at both the basic or fundamental level and advanced practical level. It is not possible to succeed in this field without the ability to perform mathematical operations.

Keep in mind that mathematics is a universal language. Mathematical symbols have the same meaning to people speaking many different languages throughout the world. The key to learning mathematics is to learn the language, symbols, definitions, and terms of mathematics that allow us to grasp the concepts necessary to solve equations.

In *Basic Mathematics for Water and Wastewater Operators*, we introduce and review concepts critical to qualified operators at the fundamental or entry level; however, this does not mean that these are the only math concepts that a competent operator must know to solve routine operation and maintenance problems. *Water Treatment Operations: Math Concepts and Calculations* and *Wastewater Treatment Operations: Math Concepts and Calculations* both present math operations that progressively advance to higher, more practical applications of mathematical calculations—that is, the math operations that operators at the highest level of licensure would be expected to know how to perform. Even though there is considerable crossover of basic math operations used by both waterworks and wastewater operators, the applied math problems for wastewater and water have been separated into separate texts. This has been done to help operators of specific unit processes unique to waterworks and wastewater operations focus on their area of specialty.

To ensure correlation to modern practice and design, we present illustrative problems in terms of commonly used waterworks/wastewater treatment operations and associated parameters and cover typical math concepts for waterworks/wastewater treatment unit process operations found in today's waterworks/wastewater treatment facilities. These texts are accessible to those who have little or no experience in treatment plant math operations. Readers who work through the texts systematically will be surprised at how easily they can acquire an understanding of water/wastewater math concepts, adding a critical component to their professional knowledge.

What makes *Water Treatment Operations: Math Concepts and Calculations* different from other math books available? Consider the following:

- The author has worked in and around water/wastewater treatment and taught water/wastewater math for several years at the apprenticeship level and at numerous short courses for operators.
- The author has sat at the table of licensure examination preparation boards to review, edit, and write state licensure exams.
- This step-by-step training manual provides concise, practical instruction in the math skills that operators must have to pass certification tests.
- The text is user friendly; no matter the difficulty of the problem to be solved, each operation is explained in straightforward, plain English. Moreover, several hundred sample problems are presented to enhance the learning process.

The original *Mathematics Manual for Water and Wastewater Treatment Plant Operators* was highly successful and well received, but like any flagship edition of any practical manual, there is always room for improvement. Many users have provided constructive criticism, advice, and numerous suggestions. All of these inputs from actual users have been incorporated into these new texts.

A final point before beginning our discussion of math concepts: It can be said with some accuracy and certainty that, without the ability to work basic math problems typical for water/wastewater treatment, candidates for licensure will find any attempts to successfully pass licensure exams a much more difficult proposition.

# Author

**Frank R. Spellman, PhD**, is a retired assistant professor of environmental health at Old Dominion University, Norfolk, Virginia, and the author of more than 90 books covering topics ranging from concentrated animal feeding operations (CAFOs) to all areas of environmental science and occupational health. Many of his texts are readily available online, and several have been adopted for classroom use at major universities throughout the United States, Canada, Europe, and Russia; two have been translated into Spanish for South American markets. Dr. Spellman has been cited in more than 450 publications. He serves as a professional expert witness for three law groups and as an incident/accident investigator for the U.S. Department of Justice and a northern Virginia law firm. In addition, he consults on homeland security vulnerability assessments for critical infrastructures, including water/wastewater facilities, and conducts audits for Occupational Safety and Health Administration and Environmental Protection Agency inspections throughout the country. Dr. Spellman receives frequent requests to co-author with well-recognized experts in various scientific fields; for example, he is a contributing author to the prestigious text *The Engineering Handbook*, 2nd ed. Dr. Spellman lectures on sewage treatment, water treatment, and homeland security, as well as on safety topics, throughout the country and teaches water/wastewater operator short courses at Virginia Tech in Blacksburg. He holds a BA in public administration, a BS in business management, an MBA, and both an MS and a PhD in environmental engineering.

# 1 Pumping Calculations

## PUMPING

Pumps and pumping calculations are discussed in detail in *Basic Mathematics for Water and Wastewater Operators*, and they are discussed here as well as in *Water Treatment Operations: Math Concepts and Calculations* because they are germane to many treatment processes, especially their influent and effluent operations. Pumping facilities and appurtenances are required wherever gravity cannot be used to supply water to the distribution system under sufficient pressure to meet all service demands. Pumps used in water and wastewater treatment are the same. Because the pump is so perfectly suited to the tasks it performs, and because the principles that make the pump work are physically fundamental, the idea that any new device would ever replace the pump is difficult to imagine. The pump is the workhorse of water/wastewater operations. Simply, pumps use energy to keep water and wastewater moving. To operate a pump efficiently, the operator or maintenance personnel must be familiar with several basic principles of hydraulics. In addition, to operate various unit processes, in both water and wastewater operations at optimum levels, operators should know how to perform basic pumping calculations.

## BASIC WATER HYDRAULICS CALCULATIONS

### Weight of Water

Because water must be stored or kept moving in water supplies and because wastewater must be collected, processed, and discharged to its receiving body, we must consider some basic relationships regarding the weight of water. One cubic foot of water weighs 62.4 lb and contains 7.48 gal. One cubic inch of water weighs 0.0362 lb. Water 1 ft deep will exert a pressure of 0.43 pounds per square inch (psi) on the bottom area (12 in. × 0.062 lb/in.$^3$). A column of water 2 ft high exerts 0.86 psi, one 10 ft high exerts 4.3 psi, and one 52 ft high exerts 52 ft × 0.43 psi/ft = 22.36 psi.

A column of water 2.31 ft high will exert 1.0 psi. To produce a pressure of 40 psi requires a water column of

$$40 \text{ psi} \times 2.31 \text{ ft/psi} = 92 \text{ ft (rounded)}$$

The term *head* is used to designate water pressure in terms of the height of a column of water in feet. For example, a 10-ft column of water exerts 4.3 psi. This can be referred to as 4.3-psi pressure or 10 ft of head. If the static pressure in a pipe leading from an elevated water storage tank is 37 psi, what is the elevation of the water above the pressure gauge? Remembering that 1 psi = 2.31 ft and that the pressure at the gauge is 37 psi, then

$$37 \text{ psi} \times 2.31 \text{ ft/psi} = 85 \text{ ft (rounded)}$$

## Weight of Water Related to the Weight of Air

The theoretical atmospheric pressure at sea level (14.7 psi) will support a column of water 34 ft high:

$$14.7 \text{ psi} \times 2.31 \text{ ft/psi} = 34 \text{ ft (rounded)}$$

At an elevation of 1 mile above sea level, where the atmospheric pressure is 12 psi, the column of water would be only 28 ft high:

$$12 \text{ psi} \times 2.31 \text{ ft/psi} = 28 \text{ ft (rounded)}$$

If a tube is placed in a body of water at sea level (e.g., glass, bucket, reservoir, lake, pool), water will rise in the tube to the same height as the water outside the tube. The atmospheric pressure of 14.7 psi will push down equally on the water surface inside and outside the tube. However, if the top of the tube is tightly capped and all of the air is removed from the sealed tube above the water surface, forming a *perfect vacuum*, then the pressure on the water surface inside the tube will be 0 psi. The atmospheric pressure of 14.7 psi on the outside of the tube will push the water up into the tube until the weight of the water exerts the same 14.7-psi pressure at a point in the tube that is even with the water surface outside the tube. The water will rise 14.7 psi × 2.31 ft/psi = 34 ft. Because it is impossible to create a perfect vacuum, the water will rise somewhat less than 34 ft; the distance it rises depends on the amount of vacuum created.

### ■ Example 1.1

*Problem:* If enough air was removed from the tube to produce an air pressure of 9.5 psi above the water in the tube, how far will the water rise in the tube?

*Solution:* To maintain the 14.7 psi at the outside water surface level, the water in the tube must produce a pressure of 14.7 psi – 9.5 psi = 5.2 psi. The height of the column of water that will produce 5.2 psi is

$$5.2 \text{ psi} \times 2.31 \text{ ft/psi} = 12 \text{ ft}$$

## Water at Rest

*Stevin's law* states: "The pressure at any point in a fluid at rest depends on the distance measured vertically to the free surface and the density of the fluid." Stated as a formula, this becomes

$$p = w \times h \tag{1.1}$$

where

$p$ = Pressure (pounds per square foot, or psf).
$w$ = Density (lb/ft$^3$).
$h$ = Vertical distance (ft).

### ■ Example 1.2

*Problem:* What is the pressure at a point 16 ft below the surface of a reservoir?

*Solution:* To calculate this, we must know that the density of water is 62.4 lb/ft$^3$. Thus,

$$p = w \times h = 62.4 \text{ lb/ft}^3 \times 16 \text{ ft} = 998.4 \text{ lb/ft}^2 \text{ (psf)}$$

Waterworks operators generally measure pressure in pounds per square *inch* rather than pounds per square *foot*; to convert, divide by 144 in.$^2$/ft$^2$ (12 in. × 12 in. = 144 in.$^2$):

$$p = \frac{998.4 \text{ lb/ft}^2}{144 \text{ in.}^2\text{/ft}^2} = 6.93 \text{ lb/in.}^2 \text{ (psi)}$$

## Gauge Pressure

We have defined head as the height a column of water would rise due to the pressure at its base. We demonstrated that a perfect vacuum plus atmospheric pressure of 14.7 psi would lift the water 34 ft. If we now open the top of the sealed tube to the atmosphere, enclose the reservoir, and increase the pressure in the reservoir, the water will again rise in the tube. Because atmospheric pressure is essentially universal, we usually ignore the first 14.7 psi of actual pressure measurements and measure only the difference between the water pressure and atmospheric pressure, the gauge pressure.

### ■ Example 1.3

*Problem:* Water in an open reservoir is subjected to 14.7 psi of atmospheric pressure, but subtracting this 14.7 psi leaves a gauge pressure of 0 psi. This shows that the water would rise 0 feet above the reservoir surface. If the gauge pressure in a water main is 110 psi, how far would the water rise in a tube connected to the main?

*Solution:*

$$110 \text{ psi} \times 2.31 \text{ ft/psi} = 254.1 \text{ ft}$$

## Water in Motion

The study of water in motion is much more complicated than that of water at rest. It is important to have an understanding of these principles because the water/wastewater in a treatment plant or distribution/collection system is nearly always in motion (much of this motion is the result of pumping, of course).

## Discharge

Discharge is the quantity of water passing a given point in a pipe or channel during a given period of time. It can be calculated by the formula:

$$Q = V \times A \tag{1.2}$$

where

$Q$ = Discharge ($ft^3$/sec, or cfs).
$V$ = Water velocity (ft/sec, or fps).
$A$ = Cross-section area of the pipe or channel ($ft^2$).

Discharge can be converted from cfs to other units such as gallons per minute (gpm) or million gallons per day (MGD) by using appropriate conversion factors.

### ■ Example 1.4

*Problem:* A pipe 12 in. in diameter has water flowing through it at 12 fps. What is the discharge in (a) cfs, (b) gpm, and (c) MGD?

*Solution:* Before we can use the basic formula, we must determine the area ($A$) of the pipe. The formula for the area is

$$A = \pi \times \frac{D^2}{4} = \pi \times r^2$$

where

$\pi$ = Constant value 3.14159.
$D$ = Diameter of the circle in feet.
$r$ = Radius of the circle in feet.

So, the area of the pipe is

$$A = \pi \times \frac{D^2}{4} = \pi \times r^2 = 3.14159 \times (0.5)^2 = 0.785 \text{ ft}^2$$

Now, for part (a), we can determine the discharge in cfs:

$$Q = V \times A = 12 \text{ fps} \times 0.785 \text{ ft}^2 = 9.42 \text{ cfs}$$

For part (b), we need to know that 1 cfs is 449 gpm, so 7.85 cfs × 449 gpm/cfs = 3520 gpm. Finally, for part (c), 1 MGD is 1.55 cfs, so

$$\frac{7.85 \text{ cfs}}{1.55 \text{ cfs/MGD}} = 5.06 \text{ MGD}$$

### The Law of Continuity

The law of continuity states that the discharge at each point in a pipe or channel is the same as the discharge at any other point (provided water does not leave or enter the pipe or channel). In equation form, this becomes

$$Q_1 = Q_2, \text{ or } A_1 \times V_1 = A_2 \times V_2 \tag{1.3}$$

#### ■ Example 1.5

*Problem:* A pipe 12 in. in diameter is connected to a 6-in.-diameter pipe. The velocity of the water in the 12-in. pipe is 3 fps. What is the velocity in the 6-in. pipe?

*Solution:* Use the equation

$$A_1 \times V_1 = A_2 \times V_2$$

to determine the area of each pipe:

- 12-inch pipe

$$A = \pi \times \frac{D^2}{4} = 3.14159 \times \frac{(1 \text{ ft})^2}{4} = 0.785 \text{ ft}^2$$

- 6-inch pipe

$$A = \pi \times \frac{D^2}{4} = 3.14159 \times \frac{(0.5 \text{ ft})^2}{4} = 0.196 \text{ ft}^2$$

The continuity equation now becomes

$$0.785 \text{ ft}^2 \times 3 \text{ fps} = 0.196 \text{ ft}^2 \times V_2$$

Solving for $V_2$,

$$V_2 = \frac{0.785 \text{ ft}^2 \times 3 \text{ fps}}{0.196 \text{ ft}^2} = 12 \text{ fps}$$

## Pipe Friction

The flow of water in pipes is caused by the pressure applied behind it either by gravity or by hydraulic machines (pumps). The flow is retarded by the friction of the water against the inside of the pipe. The resistance of flow offered by this friction depends on the size (diameter) of the pipe, the roughness of the pipe wall, and the number and type of fittings (bends, valves, etc.) along the pipe. It also depends on the speed of the water through the pipe—the more water you try to pump through

**IMPORTANT POINT**

Friction loss increases as

| | |
|---|---|
| Flow rate increases. | Pipe length increases. |
| Pipe diameter decreases. | Pipe is constricted. |
| Pipe interior becomes rougher. | Bends, fittings, and valves are added. |

a pipe, the more pressure it will take to overcome the friction. The resistance can be expressed in terms of the additional pressure required to push the water through the pipe, in either psi or feet of head. Because it is a reduction in pressure, it is often referred to as *friction loss* or *head loss*.

The actual calculation of friction loss is beyond the scope of this text. Many published tables give the friction loss in different types and diameters of pipe and standard fittings. What is more important here is recognition of the loss of pressure or head due to the friction of water flowing through a pipe. One of the factors in friction loss is the roughness of the pipe wall. A number called the *C* factor indicates pipe wall roughness; the higher the *C* factor, the smoother the pipe.

***Note:*** The term "*C* factor" is derived from the letter *C* in the Hazen–Williams equation for calculating water flow through a pipe.

Some of the roughness in the pipe will be due to the material; cast iron pipe will be rougher than plastic, for example. Additionally, the roughness will increase with corrosion of the pipe material and deposits of sediments in the pipe. New water pipes should have a *C* factor of 100 or more; older pipes can have *C* factors very much lower than this. To determine the *C* factor, we usually use published tables. In addition, when the friction losses for fittings are factored in, other published tables are available to make the proper determinations. It is standard practice to calculate the head loss from fittings by substituting the *equivalent length of pipe*, which is also available from published tables.

## BASIC PUMPING CALCULATIONS

Certain computations used for determining various pumping parameters are important to the water/wastewater operator. In this section, we cover the basic, important pumping calculations.

### Pumping Rates

***Note:*** The rate of flow produced by a pump is expressed as the volume of water pumped during a given period.

The mathematical problems most often encountered by water/wastewater operators with regard to determining pumping rates are often determined by using Equations 1.4 or 1.5:

$$\text{Pumping rate (gpm)} = \frac{\text{Gallons}}{\text{Minutes}} \tag{1.4}$$

$$\text{Pumping rate (gph)} = \frac{\text{Gallons}}{\text{Hours}} \tag{1.5}$$

### ■ Example 1.6

*Problem:* The meter on the discharge side of the pump reads in hundreds of gallons. If the meter shows a reading of 110 at 2:00 p.m. and 320 at 2:30 p.m., what is the pumping rate expressed in gallons per minute?

*Solution:* The problem asks for pumping rate in gallons per minute (gpm), so we use Equation 1.4. To solve this problem, we must first find the total gallons pumped (determined from the meter readings):

$$32{,}000 \text{ gal} - 11{,}000 \text{ gal} = 21{,}000 \text{ gal pumped}$$

The volume was pumped between 2:00 p.m. and 2:30 p.m., for a total of 30 minutes. From this information, we can calculate the gpm pumping rate:

$$\text{Pumping rate (gpm)} = \frac{21{,}000 \text{ gal}}{30 \text{ min}} = 700 \text{ gpm pumping rate}$$

### ■ Example 1.7

*Problem:* During a 15-minute pumping test, 16,000 gal were pumped into an empty rectangular tank. What is the pumping rate in gallons per minute (gpm)?

*Solution:* The problem asks for the pumping rate in gallons per minute, so again we use Equation 1.4:

$$\text{Pumping rate (gpm)} = \frac{\text{Gallons}}{\text{Minutes}} = \frac{16{,}000 \text{ gal}}{15 \text{ min}} = 1067 \text{ gpm pumping rate}$$

### ■ Example 1.8

*Problem:* A tank 50 ft in diameter is filled with water to a depth of 4 ft. To conduct a pumping test, the outlet valve to the tank is closed and the pump is allowed to discharge into the tank. After 60 minutes, the water level is 5.5 ft. What is the pumping rate in gallons per minute?

*Solution:* We must first determine the volume pumped in cubic feet:

$$\begin{aligned}\text{Volume pumped} &= \text{Area of circle} \times \text{Depth}\\ &= (0.785 \times 50 \text{ ft} \times 50 \text{ ft}) \times 1.5 \text{ ft} = 2944 \text{ ft}^3 \text{ (rounded)}\end{aligned}$$

Now convert the cubic-foot volume to gallons:

$$2944 \text{ ft}^3 \times 7.48 \text{ gal/ft}^3 = 22{,}021 \text{ gal (rounded)}$$

The pumping test was conducted over a period of 60 minutes. Using Equation 1.4, calculate the pumping rate in gallons per minute:

$$\text{Pumping rate (gpm)} = \frac{\text{Gallons}}{\text{Minutes}} = \frac{22{,}021 \text{ gal}}{60 \text{ min}} = 267 \text{ gpm (rounded) pumping rate}$$

***Note:*** Pump head measurements are used to determine the amount of energy a pump can or must impart to the water; they are measured in feet.

## Calculating Head Loss

One of the principle calculations in pumping problems is used to determine head loss. The following formula is used to calculate head loss:

$$H_f = K(V^2/2g) \tag{1.6}$$

where

$H_f$ = Friction head.
$K$ = Friction coefficient.
$V$ = Velocity in pipe.
$g$ = Gravity (32.17 ft/sec · sec).

## Calculating Head

For centrifugal pumps and positive-displacement pumps, several other important formulae are used in determining head. In centrifugal pump calculations, the conversion of the discharge pressure to discharge head is the norm. Positive-displacement pump calculations often leave given pressures in psi. In the following formulas, $W$ expresses the specific weight of liquid in pounds per cubic foot. For water at 68°F, $W$ is 62.4 lb/ft³. A water column 2.31 ft high exerts a pressure of 1 psi on 64°F water. Use the following formulas to convert discharge pressure in psig to head in feet:

- Centrifugal pump

$$H \text{ (ft)} = \frac{P \text{ (psig)} \times 2.31}{\text{Specific gravity}} \tag{1.7}$$

- Positive-displacement pump

$$H \text{ (ft)} = \frac{P \text{ (psig)} \times 144}{W} \tag{1.8}$$

To convert head into pressure:

- Centrifugal pump

$$P\text{ (psi)} = \frac{H\text{ (ft)} \times \text{Specific gravity}}{2.31} \tag{1.9}$$

- Positive-displacement pump

$$P\text{ (psi)} = \frac{H\text{ (ft)} \times W}{W} \tag{1.10}$$

## Calculating Horsepower and Efficiency

When considering work being done, we consider the "rate" at which work is being done. This is called *power* and is labeled as foot-pounds/second (ft-lb/sec). At some point in the past, it was determined that the ideal work animal, the horse, could move 550 pounds a distance of 1 foot in 1 second. Because large amounts of work are also to be considered, this unit became known as *horsepower*. When pushing a certain amount of water at a given pressure, the pump performs work. One horsepower equals 33,000 ft-lb/min. The two basic terms for horsepower are

- Hydraulic horsepower (whp)
- Brake horsepower (bhp)

### Hydraulic Horsepower

One hydraulic horsepower (whp) equals the following:

550 ft-lb/sec
33,000 ft-lb/min
2545 British thermal units per hour (Btu/hr)
0.746 kW
1.014 metric hp

To calculate the hydraulic horsepower using flow in gallons per minute and head in feet, use the following formula for centrifugal pumps:

$$\text{whp} = \frac{\text{Flow (gpm)} \times \text{Head (ft)} \times \text{Specific gravity}}{3960} \tag{1.11}$$

When calculating horsepower for positive-displacement pumps, common practice is to use psi for pressure. Then the hydraulic horsepower becomes:

$$\text{whp} = \frac{\text{Flow (gpm)} \times \text{Pressure (psi)}}{1714} \tag{1.12}$$

## Pump Efficiency and Brake Horsepower

When a motor–pump combination is used (for any purpose), neither the pump nor the motor will be 100% efficient. Not all of the power supplied by the motor to the pump (*brake horsepower*, bhp) will be used to lift the water (*water* or *hydraulic horsepower*, whp); some of the power is used to overcome friction within the pump. Similarly, not all of the power of the electric current driving the motor (*motor horsepower*, mhp) will be used to drive the pump; some of the current is used to overcome friction within the motor, and some current is lost in the conversion of electrical energy to mechanical power.

> ***Note:*** Depending on size and type, pumps are usually 50 to 85% efficient, and motors are usually 80 to 95% efficient. The efficiency of a particular motor or pump is given in the manufacturer's technical manual accompanying the unit.

The brake horsepower of a pump is equal to hydraulic horsepower divided by the pump's efficiency. Thus, the horsepower formulas become

$$\text{bhp} = \frac{\text{Flow (gpm)} \times \text{Head (ft)} \times \text{Specific gravity}}{3960 \times \text{Efficiency}} \tag{1.13}$$

or

$$\text{bhp} = \frac{\text{Flow (gpm)} \times \text{Pressure (psi)}}{1714 \times \text{Efficiency}} \tag{1.14}$$

### ■ Example 1.9

*Problem:* Calculate the bhp requirements for a pump handling saltwater and having a flow of 700 gpm with 40-psi differential pressure. The specific gravity of saltwater at 68°F equals 1.03. The pump efficiency is 85%.

*Solution:* To use Equation 1.13, convert the pressure differential to total differential head (TDH):

$$TDH = \frac{40 \times 2.31}{1.03} = 90 \text{ ft}$$

Then,

$$\text{bhp} = \frac{700 \times 90 \times 1.03}{3960 \times 0.85} = 19.3 \text{ hp (rounded)}$$

Using Equation 1.14,

$$\text{bhp} = \frac{700 \times 40}{1714 \times 0.85} = 19.2 \text{ hp (rounded)}$$

**IMPORTANT POINT**

Horsepower requirements vary with flow. Generally, if the flow is greater, the horsepower required to move the water would be greater.

When the motor, brake, and motor horsepower are known and the efficiency is unknown, a calculation to determine motor or pump efficiency must be done. Equation 1.15 is used to determine percent efficiency:

$$\text{Percent efficiency} = \frac{\text{Horsepower output}}{\text{Horsepower input}} \times 100 \tag{1.15}$$

From Equation 1.15, the specific equations to be used for motor, pump, and overall efficiency are

$$\text{Percent motor efficiency} = \frac{\text{bhp}}{\text{mhp}} \times 100$$

$$\text{Percent pump efficiency} = \frac{\text{whp}}{\text{bhp}} \times 100$$

$$\text{Percent overall efficiency} = \frac{\text{whp}}{\text{mhp}} \times 100$$

### ■ Example 1.10

*Problem:* A pump has a water horsepower requirement of 8.5 whp. If the motor supplies the pump with 10 hp, what is the efficiency of the pump?

*Solution:*

$$\text{Percent pump efficiency} = \frac{\text{whp}}{\text{bhp}} \times 100 = \frac{85\text{ whp}}{10\text{ bhp}} \times 100 = 0.85 \times 100 = 85\%$$

### ■ Example 1.11

*Problem:* What is the efficiency if an electric power equivalent to 25 hp is supplied to the motor and 16 hp of work is accomplished by the pump?

*Solution:* Calculate the percent of overall efficiency:

$$\text{Percent overall efficiency} = \frac{\text{Horsepower output}}{\text{Horsepower supplied}} \times 100$$

$$= \frac{16\text{ whp}}{25\text{ mhp}} \times 100 = 0.64 \times 100 = 64\%$$

### ■ Example 1.12

*Problem:* 12 kW of power is supplied to the motor. If the brake horsepower is 12 hp, what is the efficiency of the motor?

*Solution:* First, convert the kilowatt power to horsepower. Based on the fact that 1 hp = 0.746 kW, the equation becomes

$$\frac{12\text{ kW}}{0.746\text{ kW/hp}} = 16.09\text{ hp}$$

Now calculate the percent efficiency of the motor:

$$\text{Percent overall efficiency} = \frac{\text{Horsepower output}}{\text{Horsepower supplied}} \times 100$$

$$= \frac{12\text{ whp}}{16.09\text{ mhp}} \times 100 = 0.75 \times 100 = 75\%$$

## Specific Speed

Specific speed ($N_s$) refers to the speed of an impeller when it pumps 1 gpm of liquid at a differential head of 1 ft. Use the following equation for specific speed, where $H$ is at the best efficiency point:

$$N_s = \frac{\text{rpm} \times Q^{0.5}}{H^{0.75}} \tag{1.16}$$

where

rpm = Revolutions per minute.
$Q$ = Flow (gpm).
$H$ = Head (ft).

Pump specific speeds vary between pumps. No absolute rule sets the specific speed for different kinds of centrifugal pumps. However, the following $N_s$ ranges are quite common:

Volute, diffuser, and vertical turbine—500 to 5000
Mixed flow—5000 to 10,000
Propeller pumps—9000 to 15,000

***Note:*** The higher the specific speed of a pump, the higher its efficiency.

# POSITIVE-DISPLACEMENT PUMPS

The clearest differentiation between centrifugal (or kinetic) pumps and positive-displacement pumps can be made based on the method by which pumping energy is transmitted to the liquid. Kinetic (centrifugal pumps) pumps rely on a

**IMPORTANT POINT**

The three basic types of positive-displacement pumps discussed in this chapter are

- Reciprocating pumps
- Rotary pumps
- Special-purpose pumps (peristaltic or tubing pumps)

transformation of kinetic energy to static pressure. Positive-displacement pumps, on the other hand, discharge a given volume for each stroke or revolution; that is, energy is added intermittently to the fluid flow. The two most common forms of positive-displacement pumps are reciprocating action pumps (which use pistons, plungers, diaphragms, or bellows) and rotary action pumps (using vanes, screws, lobes, or progressing cavities). No matter which form is used, all positive-displacement pumps act to force liquid into a system regardless of the resistance that may oppose the transfer. The discharge pressure generated by a positive-displacement pump is, in theory, infinite.

## Volume of Biosolids Pumped (Capacity)

One of the most common positive-displacement biosolids pumps is the piston pump. Each stroke of a piston pump displaces, or pushes out, biosolids. Normally, the piston pump is operated at about 50 gpm. For positive-displacement pump calculations, we use the volume of biosolids pumped equation shown below:

$$\text{Biosolids pumped (gpm)} = \left[0.785 \times D^2 \times \text{Stroke length} \times 7.48 \text{ gal/ft}^3\right] \times \text{Strokes/min} \quad (1.17)$$

### ■ Example 1.13

*Problem:* A biosolids pump has a bore of 6 in. and a stroke length of 4 in. If the pump operates at 55 strokes (or revolutions) per minute, how many gpm are pumped? (Assume 100% efficiency.)

*Solution:*

$$\begin{aligned}\text{Biosolids pumped (gpm)} &= \text{Gallons pumped per stroke} \times \text{Strokes/min}\\ &= [0.785 \times D^2 \times \text{Stroke length} \times 7.48 \text{ gal/ft}^3] \times \text{Strokes/min}\\ &= [0.785 \times (0.5 \text{ ft})^2 \times 0.33 \text{ ft} \times 7.48 \text{ gal/ft}^3] \times 55 \text{ strokes/min}\\ &= 0.48 \text{ gal/stroke} \times 55 \text{ strokes/min}\\ &= 26.6 \text{ gpm}\end{aligned}$$

### ■ Example 1.14

*Problem:* A biosolids pump has a bore of 6 in. and a stroke setting of 3 in. The pump operates at 50 revolutions per minute. If the pump operates a total of 60 minutes during a 24-hour period, what is the gpd pumping rate? (Assume that the piston is 100% efficient.)

*Solution:* First calculate the gpm pumping rate:

$$\begin{aligned} \text{Biosolids pumped (gpm)} &= \text{Gallons pumped per stroke} \times \text{Strokes/min} \\ &= \left[0.785 \times (0.5\text{ ft})^2 \times 0.25\text{ ft} \times 7.48\text{ gal/ft}^3\right] \times 50\text{ strokes/min} \\ &= 0.37\text{ gal/stroke} \times 50\text{ strokes/min} \\ &= 18.5\text{ gpm} \end{aligned}$$

Then convert the gpm pumping rate to a gpd pumping rate, based on total minutes pumped during 24 hours:

$$18.5 \times 60/\text{day} = 1110\text{ gpd}$$

# 2 Water Source and Storage Calculations

## WATER SOURCES

Approximately 40 million cubic miles of water cover or reside within the Earth. The oceans contain about 97% of all water on the planet. The other 3% is freshwater, comprised of snow and ice on the surface (2.25%), usable groundwater (0.3%), and surface freshwater (0.5%). In the United States, average rainfall is approximately 2.6 feet (a volume of 5900 $km^3$). Of this amount, approximately 71% evaporates (about 4200 $km^3$), and 29% goes to stream flow (about 1700 $km^3$).

Beneficial freshwater uses include manufacturing, food production, domestic and public needs, recreation, hydroelectric power production, and flood control. Stream flow withdrawn annually is about 7.5% (440 $km^3$). Irrigation and industry use almost half of this amount (3.4%, or 200 $km^3$ per year). Municipalities use only about 0.6% (35 $km^3$ per year) of this amount.

Historically, in the United States, water usage has generally been increasing (as might be expected). In 1950, for example, freshwater withdrawals were 174 billion gallons per day. By the year 2005, that number had grown to about 349 billion gallons of freshwater being used per day.

The primary sources of freshwater include the following:

- Captured and stored rainfall in cisterns and water jars
- Groundwater from springs, artesian wells, and drilled or dug wells
- Surface water from lakes, rivers, and streams
- Desalinized seawater or brackish groundwater
- Reclaimed wastewater

In addition to the water source calculations typically used by water treatment operators in the operation of waterworks reservoirs, storage ponds, and lakes and which are included in many operator certification examinations, additional example math problems regarding algicide applications have been included in this chapter. State public water supply divisions issue permits to allow the application of copper sulfate and potassium permanganate to water supply reservoirs for algae control. Moreover, iron and manganese control via chemical treatment is, in many cases, also required. Thus, water operators must be trained in the proper pretreatment of water storage reservoirs, ponds, and lakes for the control of algae, iron, and manganese.

## WATER SOURCE CALCULATIONS

Water source calculations covered in this section apply to wells and pond/lake storage capacity. Specific well calculations discussed include well drawdown, well yield, specific yield, well casing disinfection, and deep-well turbine pump capacity.

### WELL DRAWDOWN

Drawdown is the drop in the level of water in a well when water is being pumped (see Figure 2.1). Drawdown is usually measured in feet or meters. One of the most important reasons for measuring drawdown is to make sure that the source water is adequate and not being depleted. The data collected to calculate drawdown can indicate if the water supply is slowly declining. Early detection can give the system time to explore alternative sources, establish conservation measures, or obtain any special funding that may be necessary to obtain a new water source. Well drawdown is the difference between the pumping water level and the static water level.

$$\text{Drawdown (ft)} = \text{Pumping water level (ft)} - \text{Static water level (ft)} \qquad (2.1)$$

#### ■ EXAMPLE 2.1

*Problem:* The static water level for a well is 70 ft. If the pumping water level is 110 ft, what is the drawdown?

*Solution:*

$$\text{Drawdown (ft)} = \text{Pumping water level (ft)} - \text{Static water level (ft)}$$

$$= 110 \text{ ft} - 70 \text{ ft} = 40 \text{ ft}$$

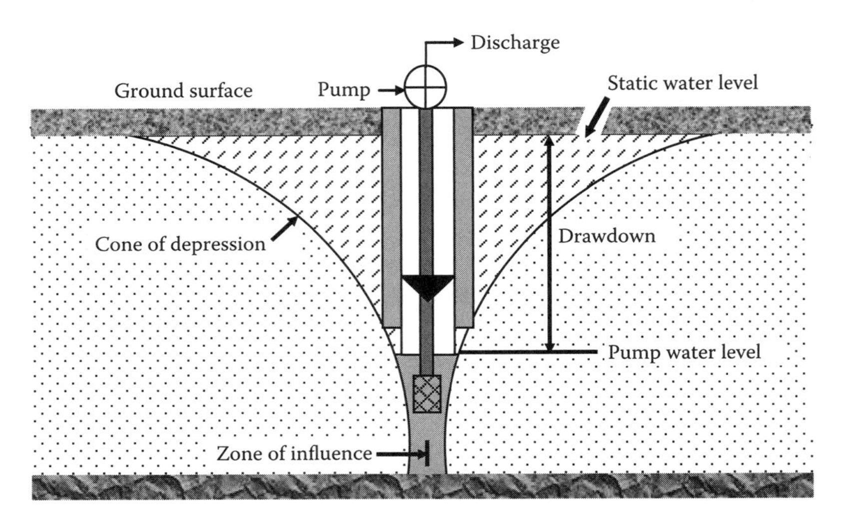


**FIGURE 2.1** Drawdown.

### ■ Example 2.2

*Problem:* The static water level of a well is 125 ft. The pumping water level is determined using a sounding line. The air pressure applied to the sounding line is 4.0 pounds per square inch (psi) and the length of the sounding line is 180 ft. What is the drawdown?

*Solution:* First calculate the water depth in the sounding line and the pumping water level:

$$\text{Water depth in sounding line} = 4.0 \text{ psi} \times 2.31 \text{ ft/psi} = 9.2 \text{ ft}$$

$$\text{Pumping water level} = 180 \text{ ft} - 9.2 \text{ ft} = 170.8 \text{ ft}$$

Then calculate drawdown as usual:

$$\text{Drawdown (ft)} = \text{Pumping water level (ft)} - \text{Static water level (ft)}$$

$$= 170.8 \text{ ft} - 125 \text{ ft} = 45.8 \text{ ft}$$

## Well Yield

Well yield is the volume of water per unit of time that is produced from the well pumping. Usually, well yield is measured in terms of gallons per minute (gpm) or gallons per hour (gph). Sometimes, large flows are measured in cubic feet per second (cfs). Well yield is determined by using the following equation:

$$\text{Well yield (gpm)} = \frac{\text{Gallons produced}}{\text{Duration of test (min)}} \tag{2.2}$$

### ■ Example 2.3

*Problem:* Once the drawdown level of a well stabilized, it was determined that the well produced 410 gal during a 5-minute test. What was the well yield?

*Solution:*

$$\text{Well yield (gpm)} = \frac{\text{Gallons produced}}{\text{Duration of test (min)}} = \frac{410 \text{ gallons}}{5 \text{ minutes}} = 82 \text{ gpm}$$

### ■ Example 2.4

*Problem:* During a 5-minute test for well yield, 760 gal are removed from the well. What is the well yield in gpm? In gph?

*Solution:* First calculate the gpm well yield:

$$\text{Well yield (gpm)} = \frac{\text{Gallons produced}}{\text{Duration of test (min)}} = \frac{760 \text{ gallons}}{5 \text{ minutes}} = 152 \text{ gpm}$$

Then convert gpm flow to gph flow:

$$152 \text{ gal/min} \times 60 \text{ min/hr} = 9120 \text{ gph}$$

## Specific Yield

Specific yield is the discharge capacity of the well per foot of drawdown. The specific yield may range from 1 gpm/ft drawdown to more than 100 gpm/ft drawdown for a properly developed well. Specific yield is calculated using the following equation:

$$\text{Specfiic yield (gpm/ft)} = \frac{\text{Well yield (gpm)}}{\text{Drawdown (ft)}} \tag{2.3}$$

### ■ Example 2.5

*Problem:* A well produces 270 gpm. If the drawdown for the well is 22 ft, what is the specific yield in gpm/ft of drawdown?

*Solution:*

$$\text{Specfiic yield} = \frac{\text{Well yield (gpm)}}{\text{Drawdown (ft)}} = \frac{270 \text{ gpm}}{22 \text{ ft}} = 12.3 \text{ gpm/ft}$$

### ■ Example 2.6

*Problem:* The yield for a particular well is 300 gpm. If the drawdown for this well is 30 ft, what is the specific yield in gpm/ft of drawdown?

*Solution:*

$$\text{Specfiic yield} = \frac{\text{Well yield (gpm)}}{\text{Drawdown (ft)}} = \frac{300 \text{ gpm}}{30 \text{ ft}} = 10 \text{ gpm/ft}$$

## Well Casing Disinfection

A new, cleaned, or a repaired well normally contains contamination, which may remain for weeks unless the well is thoroughly disinfected. This may be accomplished by ordinary bleach in a concentration of 100 parts per million (ppm) of chlorine. The amount of disinfectant required is determined by the amount of water in the well. The following equation is used to calculate the pounds of chlorine required for disinfection:

$$\text{Chlorine (lb)} = \text{Chlorine (mg/L)} \times \text{Casing volume (MG)} \times 8.34 \text{ lb/gal} \tag{2.4}$$

■ **EXAMPLE 2.7**

*Problem:* A new well is to be disinfected with chlorine at a dosage of 50 mg/L. If the well casing diameter is 8 in. and the length of the water-filled casing is 110 ft, how many pounds of chlorine will be required?

*Solution*: First calculate the volume of the water-filled casing:

$$0.785 \times (0.67)^2 \times 110 \text{ ft} \times 7.48 \text{ gal/ft}^3 = 290 \text{ gal}$$

Then determine the pounds of chlorine required using the mg/L to lb equation:

$$\text{Chlorine (mg/L)} \times \text{Casing volume (MG)} \times 8.34 \text{ lb/gal} = \text{Chlorine (lb)}$$

$$50 \text{ mg/L} \times 0.000290 \text{ MG} \times 8.34 \text{ lb/gal} = 0.12 \text{ lb chlorine}$$

## Deep-Well Turbine Pump Calculations

The deep-well turbine pump is used for high-capacity deep wells. The pump, consisting usually of more than one stage of centrifugal pump, is fastened to a pipe called the *pump column*; the pump is located in the water. The pump is driven from the surface through a shaft running inside the pump column. The water is discharged from the pump up through the pump column to the surface. The pump may be driven by a vertical shaft, electric motor at the top of the well, or some other power source, usually through a right-angle gear drive located at the top of the well. A modern version of the deep-well turbine pump is the submersible type of pump in which the pump, along with a close-coupled electric motor built as a single unit, is located below water level in the well. The motor is built to operate submerged in water.

## Vertical Turbine Pump Calculations

The calculations pertaining to well pumps include head, horsepower, and efficiency calculations. *Discharge head* is measured to the pressure gauge located close to the pump discharge flange. The pressure (psi) can be converted to feet of head using the following equation:

$$\text{Discharge head (ft)} = \text{Pressure (psi)} \times 2.31 \text{ ft/psi} \quad (2.5)$$

Total pumping head (*field head*) is a measure of the lift below the discharge head pumping water level (*discharge head*). Total pumping head is calculated as follows:

$$\text{Pumping head (ft)} = \text{Pumping water level (ft)} + \text{Discharge head (ft)} \quad (2.6)$$

### ■ Example 2.8

*Problem:* The pressure gauge reading at a pump discharge head is 4.2 psi. What is this discharge head expressed in feet?

*Solution:*

$$4.2 \text{ psi} \times 2.31 \text{ ft/psi} = 9.7 \text{ ft}$$

### ■ Example 2.9

*Problem:* The static water level of a pump is 100 ft. The well drawdown is 24 ft. If the gauge reading at the pump discharge head is 3.7 psi, what is the total pumping head?

*Solution:*

$$\text{Total pumping head (ft)} = \text{Pumping water level (ft)} + \text{Discharge head (ft)} \quad (2.7)$$

$$= (100 \text{ ft} + 24 \text{ ft}) + (3.7 \text{ psi} \times 2.31 \text{ ft/psi})$$

$$= 124 \text{ ft} + 8.5 \text{ ft} = 132.5 \text{ ft}$$

There are five types of horsepower calculations for vertical turbine pumps (refer to Figure 2.2):

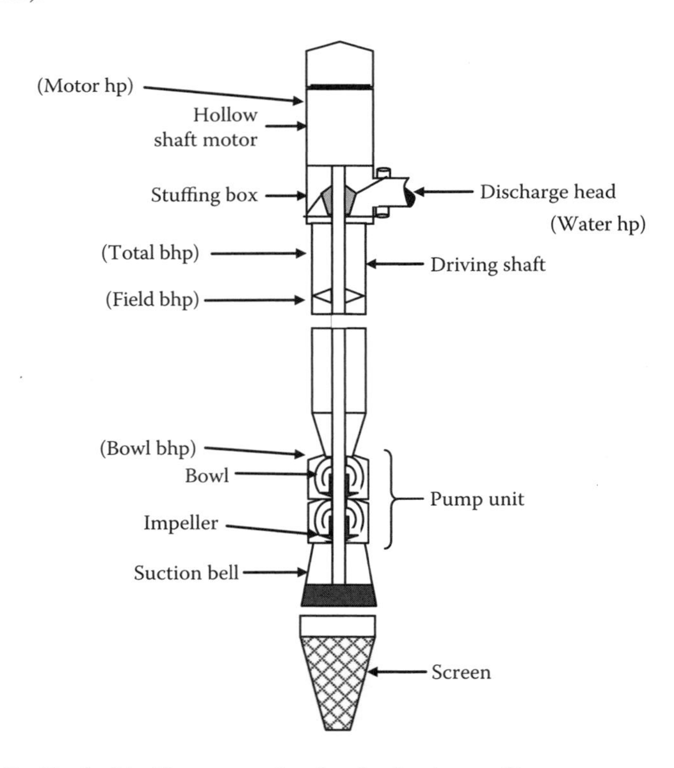


**FIGURE 2.2** Vertical turbine pump, showing the five types of horsepower.

- *Motor horsepower* refers to the horsepower supplied to the motor. The following equation is used to calculate motor horsepower:

$$\text{Motor horsepower (input hp)} = \frac{\text{Field brake horsepower (bhp)}}{\text{Motor efficiency}/100} \quad (2.8)$$

- *Total brake horsepower (bhp)* refers to the horsepower output of the motor. The following equation is used to calculate total brake horsepower:

$$\text{Total bhp} = \text{Field bhp} + \text{Thrust bearing loss (hp)} \quad (2.9)$$

- *Field brake horsepower* refers to the horsepower required at the top of the pump shaft. The following equation is used to calculate field horsepower:

$$\text{Field bhp} = \text{Bowl bhp} + \text{Shaft loss (hp)} \quad (2.10)$$

- *Bowl* or *laboratory horsepower* refers to the horsepower at the entry to the pump bowls. The following equation is used to calculate bowl horsepower:

$$\text{Bowl bhp (lab bhp)} = \frac{\text{Bowl head (ft)} \times \text{Capacity (gpm)}}{3960 \times (\text{Bowl efficiency}/100)} \quad (2.11)$$

- *Water horsepower (whp)* refers to the horsepower at the pump discharge. The following equation is used to calculate water horsepower:

$$\text{Water hp} = \frac{\text{Field head (ft)} \times \text{Capacity (gpm)}}{3960} \quad (2.12)$$

  or the equivalent equation

$$\text{Water hp} = \frac{\text{Field head (ft)} \times \text{Capacity (gpm)}}{33{,}000\ \text{ft-lb/min}}$$

### ■ Example 2.10

*Problem:* The pumping water level for a well pump is 150 ft and the discharge pressure measured at the pump discharge centerline is 3.5 psi. If the flow rate from the pump is 700 gpm, use Equation 2.12 to calculate the water horsepower.

*Solution:* First calculate the field head. The discharge head must be converted from psi to ft:

$$3.5\ \text{psi} \times 2.31\ \text{ft/psi} = 8.1\ \text{ft}$$

The field head, therefore, is

$$150\ \text{ft} + 8.1\ \text{ft} = 158.1\ \text{ft}$$

The water horsepower can now be determined:

$$\text{Water hp} = \frac{\text{Field head (ft)} \times \text{Capacity (gpm)}}{33{,}000 \text{ ft-lb/min}}$$

$$= \frac{158.1 \text{ ft} \times 700 \text{ gpm} \times 8.34 \text{ lb/gal}}{33{,}000 \text{ ft-lb/min}} = 28 \text{ whp}$$

### ■ Example 2.11

*Problem:* The pumping water level for a pump is 170 ft. The discharge pressure measured at the pump discharge head is 4.2 psi. If the pump flow rate is 800 gpm, use Equation 2.12 to calculate the water horsepower.

*Solution*: The field head must first be determined. To determine field head, the discharge head must be converted from psi to ft:

$$4.2 \text{ psi} \times 2.31 \text{ ft/psi} = 9.7 \text{ ft}$$

The field head can now be calculated:

$$170 \text{ ft} + 9.7 \text{ ft} = 179.7 \text{ ft}$$

And the water horsepower can be calculated:

$$\text{Water hp} = \frac{179.7 \text{ ft} \times 800 \text{ gpm}}{3960} = 36 \text{ whp}$$

### ■ Example 2.12

*Problem:* A deep-well vertical turbine pump delivers 600 gpm. The lab head is 185 ft and the bowl efficiency is 84%. What is the bowl horsepower?

*Solution:* Use Equation 2.11 to calculate the bowl horsepower:

$$\text{Bowl bhp} = \frac{\text{Bowl head (ft)} \times \text{Capacity (gpm)}}{3960 \times (\text{Bowl efficiency}/100)}$$

$$= \frac{185 \text{ ft} \times 600 \text{ gpm}}{3960 \times (84/100)} = \frac{185 \times 600}{3960 \times 0.84} = 33.4 \text{ bowl hp}$$

### ■ Example 2.13

*Problem:* The bowl bhp is 51.8. If the 1-in.-diameter shaft is 170 ft long and is rotating at 960 rpm with a shaft fiction loss of 0.29 hp loss per 100 ft, what is the field bhp?

*Solution:* Before field bhp can be calculated, the shaft loss must be factored in:

$$\frac{0.29 \text{ hp loss} \times 170 \text{ ft}}{100} = 0.5 \text{ hp loss}$$

Now the field bhp can be determined:

$$\text{Field bhp} = \text{Bowl bhp} + \text{Shaft loss (hp)} = 51.8 + 0.5 = 52.3 \text{ bhp}$$

## ■ Example 2.14

*Problem:* The field horsepower for a deep-well turbine pump is 62 bhp. If the thrust bearing loss is 0.5 hp and the motor efficiency is 88%, what is the motor input horsepower?

*Solution:*

$$\text{Motor hp} = \frac{\text{Total bhp}}{\text{Motor efficiency/100}} = \frac{62 \text{ bhp} + 0.5 \text{ hp}}{0.88} = 71 \text{ mhp}$$

When we speak of the *efficiency* of any machine, we are speaking primarily of a comparison of what is put out by the machine (e.g., energy output) compared to its input (e.g., energy input). Horsepower efficiency, for example, is a comparison of horsepower output of the unit or system with horsepower input to that unit or system—the unit's efficiency. There are four types efficiencies considered with vertical turbine pumps:

- Bowl efficiency
- Field efficiency
- Motor efficiency
- Overall efficiency

The general equation used in calculating percent efficiency is shown below:

$$\text{Percent (\%)} = \frac{\text{Part}}{\text{Whole}} \times 100 \tag{2.13}$$

Vertical turbine pump bowl efficiency is easily determined using a pump performance curve chart provided by the pump manufacturer.

Field efficiency is determined by

$$\text{Field efficiency (\%)} = \frac{\left[\text{Field head (ft)} \times \text{Capacity (gpm)}\right]/3960}{\text{Total bhp}} \times 100 \tag{2.14}$$

### ■ Example 2.15

*Problem:* Given the data below, calculate the field efficiency of the deep-well turbine pump:

Field head = 180 ft
Capacity = 850 gpm
Total bhp = 61.3 bhp

*Solution:*

$$\text{Field efficiency (\%)} = \frac{\text{Field head (ft)} \times \text{Capacity (gpm)}}{3960 \times \text{Total bhp}} \times 100$$

$$= \frac{180 \text{ ft} \times 850 \text{ gpm}}{3960 \times 61.3 \text{ bhp}} \times 100 = 63\%$$

The *overall efficiency* is a comparison of the horsepower output of the system with that entering the system. Equation 2.15 is used to calculate overall efficiency:

$$\text{Overall efficiency (\%)} = \frac{\text{Field efficiency (\%)} \times \text{Motor efficiency (\%)}}{100} \tag{2.15}$$

### ■ Example 2.16

*Problem:* The efficiency of a motor is 90%. If the field efficiency is 83%, what is the overall efficiency of the unit?

*Solution:*

$$\text{Overall efficiency} = \frac{\text{Field efficiency (\%)} \times \text{Motor efficiency (\%)}}{100} = \frac{83 \times 90}{100} = 74.7\%$$

## WATER STORAGE

Water storage facilities for water distribution systems are required primarily to provide for fluctuating demands of water usage (to provide a sufficient amount of water to average or equalize daily demands on the water supply system). In addition, other functions of water storage facilities include increasing operating convenience, leveling pumping requirements (to keep pumps from running 24 hours a day), decreasing power costs, providing water during power source or pump failure, providing large quantities of water to meet fire demands, providing surge relief (to reduce the surge associated with stopping and starting pumps), increasing detention time (to provide chlorine contact time and satisfy the desired contact time requirements), and blending water sources.

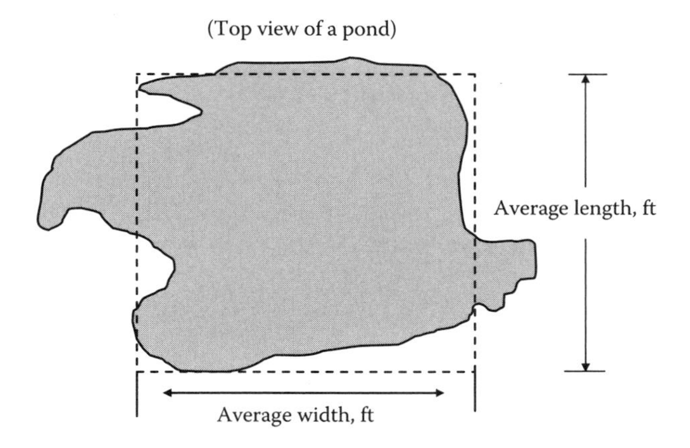


**FIGURE 2.3** Determining pond storage capacity.

## WATER STORAGE CALCULATIONS

The storage capacity, in gallons, of a reservoir, pond, or small lake can be estimated (see Figure 2.3) using Equation 2.16:

$$\text{Capacity} = \text{Avg. length (ft)} \times \text{Avg. width (ft)} \times \text{Avg. depth (ft)} \times 7.48\ \text{gal/ft}^3 \tag{2.16}$$

### ■ Example 2.17

*Problem:* A pond has an average length of 250 ft, an average width of 110 ft, and an estimated average depth of 15 ft. What is the estimated volume of the pond in gallons?

*Solution:*

$$\text{Volume} = \text{Avg. length (ft)} \times \text{Avg. width (ft)} \times \text{Avg. depth (ft)} \times 7.48\ \text{gal/ft}^3$$

$$= 250\ \text{ft} \times 110\ \text{ft} \times 15\ \text{ft} \times 7.48\ \text{gal/ft}^3 = 3{,}085{,}500\ \text{gal}$$

### ■ Example 2.18

*Problem:* A small lake has an average length of 300 ft and an average width of 95 ft. If the maximum depth of the lake is 22 ft, what is the estimated volume of the lake in gallons?

> ***Note:*** For small ponds and lakes, the average depth is generally about 0.4 times the greatest depth; therefore, to estimate the average depth, measure the greatest depth, then multiply that number by 0.4.

*Solution*: First, the average depth of the lake must be estimated:

$$\text{Estimated average depth (ft)} = \text{Greatest depth (ft)} \times 0.4 \text{ ft} = 22 \text{ ft} \times 0.4 \text{ ft} = 8.8 \text{ ft}$$

Then the lake volume can be determined:

$$\text{Volume} = \text{Avg. length (ft)} \times \text{Avg. width (ft)} \times \text{Avg. depth (ft)} \times 7.48 \text{ gal/ft}^3$$

$$= 300 \text{ ft} \times 95 \text{ ft} \times 8.8 \text{ ft} \times 7.48 \text{ gal/ft}^3 = 1{,}875{,}984 \text{ gal}$$

## COPPER SULFATE DOSING

Algae control by applying copper sulfate is perhaps the most common *in situ* treatment of lakes, ponds, and reservoirs; the copper ions in the water kill the algae. Copper sulfate application methods and dosages will vary depending on the specific surface water body being treated, and they also depend on alkalinity and pH. For example, if methyl-orange alkalinity is <50 mg/L as calcium carbonate ($CaCO_3$), then a dosage of 0.3 mg/L is recommended, based on total lake or reservoir volume. On the other hand, if methyl-orange alkalinity is >50 mg/L as $CaCO_3$, 1 mg/L for the upper 2 feet of the volume of the lake or reservoir is recommended. The desired copper sulfate dosage may be expressed in mg/L copper, lb/ac-ft copper sulfate, or lb/ac copper sulfate. Potassium permanganate ($KMnO_4$) is added to ponds and reservoirs to oxidize iron (Fe) and manganese (Mn), and may be helpful in controlling algae.

For a dose expressed as mg/L copper, the following equation is used to calculate the pounds of copper sulfate required:

$$\text{Copper sulfate (lb)} = \frac{\text{Copper (mg/L)} \times \text{Volume (MG)} \times 8.34 \text{ lb/gal}}{\text{\% Available copper/100}} \tag{2.17}$$

### ■ Example 2.19

*Problem:* For algae control in a small pond, a dosage of 0.5 mg/L copper is desired. The pond has a volume of 15 MG. How many pounds of copper sulfate will be required? (Copper sulfate contains 25% available copper.)

*Solution:*

$$\text{Copper sulfate} = \frac{\text{Copper (mg/L)} \times \text{Volume (MG)} \times 8.34 \text{ lb/gal}}{\text{\% Available copper/100}}$$

$$= \frac{0.5 \text{ mg/L} \times 15 \text{ MG} \times 8.34 \text{ lb/gal}}{25/100} = 250 \text{ lb}$$

For calculating lb/ac-ft copper sulfate, use the following equation to determine the pounds of copper sulfate required (assume the desired copper sulfate dosage is 0.9 lb/ac-ft):

$$\text{Copper sulfate (lb)} = \frac{0.9 \text{ lb copper sulfate} \times \text{ac-ft}}{1 \text{ ac-ft}} \tag{2.18}$$

## ■ EXAMPLE 2.20

*Problem:* A pond has a volume of 35 ac-ft. If the desired copper sulfate dose is 0.9 lb/ac-ft, how many pounds of copper sulfate will be required?

*Solution:*

$$\text{Copper sulfate (lb)} = \frac{0.9 \text{ lb/ac-ft copper sulfate} \times \text{ac-ft}}{1 \text{ ac-ft}}$$

$$\frac{0.9 \text{ lb/ac-ft copper sulfate}}{1 \text{ ac-ft}} = \frac{x \text{ lb copper sulfate}}{35 \text{ ac-ft}}$$

Then solve for $x$:

$$0.9 \text{ lb/ac-ft} \times 35 \text{ ac-ft} = x \text{ lb}$$

$$x = 31.5 \text{ lb}$$

The desired copper sulfate dosage may also be expressed in terms of lb/ac copper sulfate. The following equation is used to determine the pounds of copper sulfate required (assume the desired copper sulfate dosage is 5.2 lb/ac):

$$\text{Copper sulfate (lb)} = \frac{5.2 \text{ lb copper sulfate} \times \text{ac}}{1 \text{ ac}} \tag{2.19}$$

## ■ EXAMPLE 2.21

*Problem:* A small lake has a surface area of 6 ac. If the desired copper sulfate dose is 5.2 lb/ac, how many pounds of copper sulfate are required?

*Solution:*

$$\text{Copper sulfate} = \frac{5.2 \text{ lb/ac copper sulfate} \times 6 \text{ ac}}{1 \text{ ac}} = 31.2 \text{ lb}$$

## ■ EXAMPLE 2.22

*Problem 1*: A holding pond measures 500 ft by 1220 ft and has an average depth of 11 ft. What is the volume of the pond in ac-ft?

*Solution:*

$$\text{Area} = \frac{\text{Length (ft)} \times \text{Width (ft)}}{43{,}560\ \text{ft}^2/\text{ac}} = \frac{500\ \text{ft} \times 1220\ \text{ft}}{43{,}560\ \text{ft}^2/\text{ac}} = 14\ \text{ac}$$

$$\text{Volume (ac-ft)} = 14\ \text{ac} \times 11\ \text{ft} = 154\ \text{ac-ft}$$

*Problem 2*: What is the volume of the pond in million gallons?

*Solution:*

$$\begin{aligned}\text{Volume (gal)} &= \text{Volume (ac-ft)} \times 43{,}560\ \text{ft}^2/\text{ac} \times 7.48\ \text{gal/ft}^3\\ &= 154\ \text{ac-ft} \times 43{,}560\ \text{ft}^2/\text{ac} \times 7.48\ \text{gal/ft}^3\\ &= 50{,}177{,}635\ \text{gal}/(1{,}000{,}000\ \text{gal/MG}) = 50.18\ \text{MG}\end{aligned}$$

### ■ Example 2.23

*Problem:* For algae control of a reservoir, a dosage of 0.6 mg/L copper is desired. The reservoir has a volume of 21 MG. How many pounds of copper sulfate (25% available copper) will be required?

*Solution:*

$$\begin{aligned}\text{Copper sulfate} &= \frac{\text{Dose (mg/L)} \times \text{Volume (MG)} \times 8.34\ \text{lb/gal}}{\%\ \text{Copper}}\\ &= \frac{0.6\ \text{mg/L} \times 21\ \text{MG} \times 8.34\ \text{lb/gal}}{0.25} = 420.3\ \text{lb}\end{aligned}$$

### ■ Example 2.24

*Problem:* The desired copper sulfate does in a reservoir is 4 mg/L. The reservoir has a volume of 60 ac-ft. How many pounds of copper sulfate (25% available copper) will be required?

*Solution:*

$$60\ \text{ac-ft} \times \left(\frac{325{,}828.8}{1\ \text{ac-ft}}\right) \times \left(\frac{1\ \text{MG}}{1{,}000{,}000\ \text{gal}}\right) = 19.5\ \text{MG}$$

$$\text{Copper sulfate} = \frac{4\ \text{mg/L} \times 19.5\ \text{MG} \times 8.34\ \text{lb/gal}}{0.25} = 2602.1\ \text{lb}$$

### ■ Example 2.25

*Problem:* A pond has an average length of 260 ft, an average width of 80 ft, and an average depth of 12 ft. If the desired dose of copper sulfate is 0.9 lb/ac-ft, how many pounds of copper sulfate will be required?

*Solution:*

$$\text{Volume (ac-ft)} = \frac{260 \text{ ft} \times 80 \text{ ft} \times 12 \text{ ft}}{43{,}560 \text{ ft}^2/\text{ac}} = 5.7 \text{ ac-ft}$$

$$\text{Copper sulfate} = 5.7 \text{ ac-ft} \times 0.9 \text{ lb/ac-ft} = 5.13 \text{ lb}$$

## IRON AND MANGANESE REMOVAL

Iron and manganese are frequently found in groundwater and in some surface waters. They do not cause health-related problems but are objectionable because they may cause aesthetic problems that cause consumers to avoid an otherwise safe water supply. Consumers might turn to a water supply of unknown or questionable quality, or they may incur unnecessary expenses for bottled water. Aesthetic problems associated with iron and manganese include the discoloration of water (iron, reddish water; manganese, brown or black water); staining of plumbing fixtures; imparting a bitter taste to the water; and stimulating the growth of microorganisms. Although no direct health concerns are associated with iron and manganese, the growth of iron bacteria slimes may cause indirect health problems. Economic problems include damage to textiles, dye, paper, and food. Iron residue (or tuberculation) in pipes increases pumping head, decreases carrying capacity, may clog pipes, and may corrode through pipes.

***Note:*** Iron and manganese are secondary contaminants. Their secondary maximum contaminant levels (SMCLs) are 0.3 mg/L for iron and 0.05 mg/L for manganese.

Iron and manganese are most likely found in groundwater supplies, industrial waste, and acid mine drainage and as byproducts of pipeline corrosion. They may accumulate in lake and reservoir sediments, causing possible problems during lake or reservoir turnover. They are not usually found in running waters (streams, rivers, etc.).

The chemical precipitation treatments for iron and manganese removal are *deferrization* and *demanganization*. The usual process is *aeration*; dissolved oxygen is the chemical causing precipitation, and chlorine or potassium permanganate may be required.

### Precipitation

Precipitation (or pH adjustment) or iron and manganese from water in their solid forms can be effected in treatment plants by adjusting the pH of the water by adding lime or other chemicals. Some of the precipitate will settle out with time, while the rest is easily removed by sand filters. This process requires the pH of the water to be in the range of 10 to 11.

***Note:*** Although the precipitation or pH adjustment technique for treating water containing iron and manganese is effective, note that the pH level must be adjusted higher (10 to 11) to cause the precipitation, which means that the pH level must then also be lowered (to 8.5 or a bit lower) to use the water for consumption.

## Oxidation

One of the most common methods of removing iron and manganese is through the process of oxidation (another chemical process), usually followed by settling and filtration. Air, chlorine, or potassium permanganate can oxidize these minerals. Each oxidant has advantages and disadvantages, and each operates slightly differently:

1. *Air*—To be effective as an oxidant, the air must come in contact with as much of the water as possible. Aeration is often accomplished by bubbling diffused air through the water by spraying the water up into the air, or by trickling the water over rocks, boards, or plastic packing materials in an aeration tower. The more finely divided the drops of water, the more oxygen comes in contact with the water and the dissolved iron and manganese.
2. *Chlorine*—This is one of the most popular oxidants for iron and manganese control because it is also widely used as a disinfectant; iron and manganese control by prechlorination can be as simple as adding a new chlorine feed point in a facility already feeding chlorine. It also provides a pre-disinfecting step that can help control bacterial growth through the rest of the treatment system. The downside to the use of chorine, however, is that when chlorine reacts with the organic materials found in surface water and some groundwaters, it forms trihalomethanes (TTHMs). This process also requires that the pH of the water be in the range of 6.5 to 7; because many groundwaters are more acidic than this, pH adjustment with lime, soda ash, or caustic soda may be necessary when oxidizing with chlorine.
3. *Potassium permanganate*—This is the best oxidizing chemical to use for manganese control removal. An extremely strong oxidant, it has the additional benefit of producing manganese dioxide during the oxidation reaction. Manganese dioxide acts as an adsorbent for soluble manganese ions. This attraction for soluble manganese provides removal to extremely low levels.

The oxidized compounds form precipitates that are removed by a filter. Note that sufficient time should be allowed from the addition of the oxidant to the filtration step; otherwise, the oxidation process will be completed after filtration, creating insoluble iron and manganese precipitates in the distribution system.

## Ion Exchange

The ion exchange process is used primarily to soften hard waters, but it will also remove soluble iron and manganese. The water passes through a bed of resin that adsorbs undesirable ions from the water, replacing them with less troublesome ions. When the resin has given up all of its donor ions, it is regenerated with strong salt brine (sodium chloride); the sodium ions from the brine replace the adsorbed ions and restore the ion exchange capabilities.

### Sequestering

Sequestering or stabilization may be used when the water contains mainly low concentrations of iron and the volumes needed are relatively small. This process does not actually remove the iron or manganese from the water but complexes it (binds it chemically) with other ions in a soluble form that is not likely to come out of solution (i.e., not likely to be oxidized).

### Aeration

The primary physical process uses air to oxidize the iron and manganese. The water is either pumped up into the air or allowed to fall over an aeration device. The air oxidizes the iron and manganese which is then removed by use of a filter. The addition of lime to raise the pH is often added to the process. Although this is called a physical process, removal is accomplished by chemical oxidation.

### Potassium Permanganate Oxidation and Manganese Greensand

The continuous regeneration potassium permanganate greensand filter process is another commonly used filtration technique for iron and manganese control. Manganese greensand is a mineral (gluconite) that has been treated with solutions of manganous chloride and potassium permanganate. The result is a sand-like (zeolite) material coated with a layer of manganese dioxide—an adsorbent for soluble iron and manganese. Manganese greensand has the ability to capture (adsorb) soluble iron and manganese that may have escaped oxidation, as well as the capability of physically filtering out the particles of oxidized iron and manganese. Manganese greensand filters are generally set up as pressure filters, totally enclosed tanks containing the greensand. The process of adsorbing soluble iron and manganese "uses up" the greensand by converting the manganese dioxide coating to manganic oxide, which does not have the adsorption property. The greensand can be regenerated in much the same way as ion exchange resins, by washing the sand with potassium permanganate.

#### ■ Example 2.26

*Problem:* A chemical supplier recommends a 3% permanganate solution. If 2 lb $KMnO_4$ are dissolved in 12 gal of water, what is the percent by weight?

*Solution:*

$$12 \text{ gal} \times 8.34 \text{ lb/gal} = 101 \text{ lb}$$

$$\text{Percent strength} = \frac{2 \text{ lb}}{101 \text{ lb} + 2 \text{ lb}} \times 100 = 1.94\%$$

#### ■ Example 2.27

*Problem:* To produce a 3% solution, how many pounds $KMnO_4$ should be dissolved in a tank 4 ft in diameter and filled to a depth of 4 ft?

*Solution:*

$$\text{Volume (gal)} = 0.785 \times 4 \times 4 \times 4 \times 7.48 = 375.8 \text{ gal}$$

$$\text{Chemical (lb)} = \frac{\text{Water volume} \times 8.34 \times \% \text{ Solution}}{1 - \% \text{ Solution}}$$

$$= \frac{375.8 \text{ gal} \times 8.34 \times \% \text{ Solution}}{1 - 0.03} = 96.94 \text{ lb}$$

■ **Example 2.28**

*Problem:* A plant's raw water has 1.6 mg/L of iron. How much $KMnO_4$ should be used to treat the iron? (Each 1.0 ppm of iron requires 0.91 mg/L of $KMnO_4$.)

*Solution:*

$$1.6 \text{ mg/L iron} \times \left( \frac{0.91 \text{ mg/L } KMnO_4}{1 \text{ ppm Fe}} \right) = 1.46 \text{ mg/L}$$

■ **Example 2.29**

*Problem:* A plant's raw water has 6.8 mg/L of manganese. How much $KMnO_4$ should be used to treat the manganese? (Each 1 ppm of manganese requires 1.92 mg/L of $KMnO_4$.)

*Solution:*

$$KMnO_4 = 6.8 \text{ mg/L} \times 1.92 \text{ mg/L} = 13.06 \text{ mg/L}$$

■ **Example 2.30**

*Problem:* A plant's raw water has 0.3 mg/L of iron and 2.6 mg/L of manganese. How much $KMnO_4$ should be used to treat them? (Assume 0.91 mg/L $KMnO_4$ per 1 ppm iron is required, and 1.92 mg/L $KMnO_4$ per 1 ppm manganese.)

*Solution:*

$$\text{For the iron, } KMnO_4 = 0.3 \text{ mg/L} \times 0.91 \text{ mg/L} = 0.273 \text{ mg/L}$$

$$\text{For the manganese, } KMnO_4 = 2.6 \text{ mg/L} \times 1.92 \text{ mg/L} = 4.99 \text{ mg/L}$$

$$\text{Total } KMnO_4 = 0.273 + 4.99 = 5.26 \text{ mg/L}$$

■ **Example 2.31**

*Problem:* A chemical supplier recommends a 3% permanganate solution mixed at a ratio of 0.22 lb per 1 gallon of water. How many milligrams $KMnO_4$ are there per milliliter of solution?

*Solution:*

$$\frac{0.22 \text{ lb}}{\text{gal}} \times \frac{1 \text{ gal}}{3785 \text{ mL}} \times \frac{453.69 \text{ g}}{1 \text{ lb}} \times \frac{1000 \text{ mg}}{1 \text{ g}} = 26.4 \text{ mg/mL}$$

### ■ Example 2.32

*Problem:* A chemical supplier recommends a 3% permanganate solution mixed at a ratio of 0.25 lb per 1 gallon of water. If 60 gal of $KMnO_4$ are made at this ratio, how many pounds of chemical are required?

*Solution:*

$$KMnO_4 = 0.25 \text{ lb/gal} \times 60 \text{ gal} = 15 \text{ lb}$$

# 3 Coagulation and Flocculation Calculations

## COAGULATION

Following screening and the other pretreatment processes, the next unit process in a conventional water treatment system is a mixer, where the first chemicals are added in what is known as coagulation. The exception to this situation occurs in small systems using groundwater, when chlorine or other taste and odor control measures are introduced at the intake and are the extent of treatment. The process of coagulation is a series of chemical and mechanical operations by which coagulants are applied and made effective. These operations are comprised of two distinct phases: (1) rapid mixing to disperse coagulant chemicals by violent agitation into the water being treated, and (2) flocculation to agglomerate small particles into well-defined floc by gentle agitation for a much longer time. The coagulant must be added to the raw water and perfectly distributed into the liquid; such uniformity of chemical treatment is reached through rapid agitation or mixing. Coagulation results from adding salts of iron or aluminum to the water and is a reaction between one of the following (coagulants) salts and water:

- Aluminum sulfate (alum)
- Sodium aluminate
- Ferric sulfate
- Ferrous sulfate
- Ferric chloride
- Polymers

## FLOCCULATION

Flocculation follows coagulation in the conventional water treatment process. Flocculation is the physical process of slowly mixing the coagulated water to increase the probability of particle collision. Through experience, we see that effective mixing reduces the required amount of chemicals and greatly improves the sedimentation process, which results in longer filter runs and higher quality finished water. The goal of flocculation is to form a uniform, feather-like material similar to snowflakes—a dense, tenacious floc that traps the fine, suspended, and colloidal particles and carries them down rapidly in the settling basin. To increase the speed of floc formation and the strength and weight of the floc, polymers are often added.

## COAGULATION AND FLOCCULATION CALCULATIONS

In the proper operation of the coagulation and flocculation unit processes, calculations are performed to determine chamber or basin volume, chemical feed calibration, chemical feeder settings, and detention time.

### Chamber and Basin Volume Calculations

To determine the volume of a square or rectangular chamber or basin, we use Equation 3.1 or Equation 3.2:

$$\text{Volume (ft}^3) = \text{Length (ft)} \times \text{Width (ft)} \times \text{Depth (ft)} \tag{3.1}$$

$$\text{Volume (gal)} = \text{Length (ft)} \times \text{Width (ft)} \times \text{Depth (ft)} \times 7.48 \text{ gal/ft}^3 \tag{3.2}$$

#### ■ Example 3.1

*Problem:* A flash mix chamber is 4 ft square with water to a depth of 3 ft. What is the volume of water (in gallons) in the chamber?

*Solution:*

$$\text{Chamber volume} = \text{Length (ft)} \times \text{Width (ft)} \times \text{Depth (ft)} \times 7.48 \text{ gal/ft}^3$$

$$= 4 \text{ ft} \times 4 \text{ ft} \times 3 \text{ ft} \times 7.48 \text{ gal/ft}^3 = 359 \text{ gal}$$

#### ■ Example 3.2

*Problem:* A flocculation basin is 40 ft long by 12 ft wide, with water to a depth of 9 ft. What is the volume of water (in gallons) in the basin?

*Solution:*

$$\text{Basin volume} = \text{Length (ft)} \times \text{Width (ft)} \times \text{Depth (ft)} \times 7.48 \text{ gal/ft}^3$$

$$= 40 \text{ ft} \times 12 \text{ ft} \times 9 \text{ ft} \times 7.48 \text{ gal/ft}^3 = 32{,}314 \text{ gal}$$

#### ■ Example 3.3

*Problem:* A flocculation basin is 50 ft long by 22 ft wide and contains water to a depth of 11 ft, 6 in. How many gallons of water are in the tank?

*Solution:* First convert the 6-inch portion of the depth measurement to feet:

$$(6 \text{ in.}) \div (12 \text{ in./ft}) = 0.5 \text{ ft}$$

Then use Equation 3.2 to calculate basin volume:

$$\text{Basin volume} = \text{Length (ft)} \times \text{Width (ft)} \times \text{Depth (ft)} \times 7.48 \text{ gal/ft}^3$$

$$= 50 \text{ ft} \times 22 \text{ ft} \times 11.5 \text{ ft} \times 7.48 \text{ gal/ft}^3 = 94{,}622 \text{ gal}$$

## Detention Time

Because coagulation reactions are rapid, detention time for flash mixers is measured in seconds, whereas the detention time for flocculation basins is generally between 5 and 30 minutes. The equation used to calculate detention time is shown below:

$$\text{Detention time (min)} = \frac{\text{Volume of tank (gal)}}{\text{Flow rate (gpm)}} \tag{3.3}$$

### ■ Example 3.4

*Problem:* The flow to a flocculation basin 50 ft long by 12 ft wide and 10 ft deep is 2100 gpm. What is the detention time in the tank in minutes?

*Solution:*

$$\text{Tank volume} = 50\text{ ft} \times 12\text{ ft} \times 10\text{ ft} \times 7.48\text{ gal/ft}^3 = 44{,}880\text{ gal}$$

$$\text{Detention time} = \frac{\text{Volume of tank (gal)}}{\text{Flow rate (gpm)}} = \frac{44{,}880\text{ gal}}{2100\text{ gpm}} = 21.4\text{ min}$$

### ■ Example 3.5

*Problem:* Assume the flow is steady and continuous for a flash mix chamber that is 6 ft long by 4 ft wide, with water to a depth of 3 ft. If the flow to the flash mix chamber is 6 million gallons per day (MGD), what is the chamber detention time in seconds?

*Solution:* First, convert the flow rate from gallons per day (gpd) to gallons per second (gps) so the time units will match:

$$\frac{6{,}000{,}000}{1440\text{ min/day} \times 60\text{ sec/min}} = 69\text{ gps}$$

Then calculate detention time using Equation 3.3:

$$\text{Detention time} = \frac{\text{Volume of tank (gal)}}{\text{Flow rate (gps)}} = \frac{6\text{ ft} \times 4\text{ ft} \times 3\text{ ft} \times 7.48\text{ gal/ft}^3}{69\text{ gps}} = 7.8\text{ sec}$$

## Determining Dry Chemical Feeder Setting

When adding (dosing) chemicals to the water flow, a measured amount of chemical is called for. The amount of chemical required depends on such factors as the type of chemical used, the reason for dosing, and the flow rate being treated. To convert from mg/L to lb/day, the following equation is used:

$$\text{Chemical added (lb/day)} = \text{Chemical (mg/L)} \times \text{Flow (MGD)} \times 8.34\text{ lb/gal} \tag{3.4}$$

### ■ Example 3.6

*Problem:* Jar tests indicate that the best alum dose for a water is 8 mg/L. If the flow to be treated is 2,100,000 gpd, what should be the lb/day setting on the dry alum feeder?

*Solution:*

$$\text{Chemical added} = \text{Chemical (mg/L)} \times \text{Flow (MGD)} \times 8.34 \text{ lb/gal}$$

$$= 8 \text{ mg/L} \times 2.10 \text{ MGD} \times 8.34 \text{ lb/gal} = 140 \text{ lb/day}$$

### ■ Example 3.7

*Problem:* Determine the desired lb/day setting on a dry chemical feeder if jar tests indicate an optimum polymer dose of 12 mg/L and the flow to be treated is 4.15 MGD.

*Solution:*

$$\text{Polymer} = 12 \text{ mg/L} \times 4.15 \text{ MGD} \times 8.34 \text{ lb/gal} = 415 \text{ lb/day}$$

## Determining Chemical Solution Feeder Setting (lb/day)

When solution concentration is expressed as pounds of chemical per gallon of solution, the required feed rate can be determined using the following equations:

$$\text{Chemical (lb/day)} = \text{Chemical (mg/L)} \times \text{Flow (MGD)} \times 8.34 \text{ lb/gal} \quad (3.5)$$

Then convert the lb/day dry chemical to a gpd solution:

$$\text{Solution (gpd)} = \frac{\text{Chemical (lb/day)}}{\text{Pounds chemical per gallon solution}} \quad (3.6)$$

### ■ Example 3.8

*Problem:* Jar tests indicate that the best alum dose is 7 mg/L. The volume to be treated is 1.52 MGD. Determine the gpd setting for the alum solution feeder if the liquid alum contains 5.36 lb of alum per gallon of solution.

*Solution:* First calculate the lb/day of dry alum required, using the mg/L to lb/day equation:

$$\text{Dry alum} = \text{Chemical (mg/L)} \times \text{Flow (MGD)} \times 8.34 \text{ lb/gal}$$

$$= 7 \text{ mg/L} \times 1.52 \text{ MGD} \times 8.34 \text{ lb/gal} = 89 \text{ lb/day}$$

Then calculate the gpd solution required:

$$\text{Alum solution} = \frac{89 \text{ lb/day}}{5.36 \text{ lb alum per gallon solution}} = 16.6 \text{ gpd}$$

### Determining Chemical Solution Feeder Setting (mL/min)

Some solution chemical feeders dispense chemical as milliliters per minute (mL/min). To calculate the mL/min feed rate required, use the following equation:

$$\text{Feed rate (mL/min)} = \frac{\text{Feed rate (gpd)} \times 3785 \text{ mL/gal}}{1440 \text{ min/day}} \tag{3.7}$$

#### ■ Example 3.9

*Problem:* The desired solution feed rate was calculated to be 9 gpd. What is this feed rate expressed as mL/min?

*Solution:*

$$\text{Feed rate (mL/min)} = \frac{\text{Feed rate (gpd)} \times 3785 \text{ mL/gal}}{1440 \text{ min/day}}$$

$$= \frac{9 \text{ gpd} \times 3785 \text{ mL/gal}}{1440 \text{ min/day}} = 24 \text{ mL/min}$$

#### ■ Example 3.10

*Problem:* The desired solution feed rate has been calculated to be 25 gpd. What is this feed rate expressed as mL/min?

*Solution:*

$$\text{Feed rate (mL/min)} = \frac{\text{Feed rate (gpd)} \times 3785 \text{ mL/gal}}{1440 \text{ min/day}}$$

$$= \frac{25 \text{ gpd} \times 3785 \text{ mL/gal}}{1440 \text{ min/day}} = 65.7 \text{ mL/min}$$

Sometimes we need to know the mL/min solution feed rate but we do not know the gpd solution feed rate. In such cases, calculate the gpd solution feed rate first, using the following the equation:

$$\text{Feed rate (gpd)} = \frac{\text{Chemical (mg/L)} \times \text{Flow (MGD)} \times 8.34 \text{ lb/gal}}{\text{Pounds chemical per gallon solution}} \tag{3.8}$$

## Determining Percent Strength of Solutions

The strength of a solution is a measure of the amount of chemical solute dissolved in the solution. We use the following equation to determine the percent strength of a solution:

$$\text{Percent strength} = \frac{\text{Chemical (lb)}}{\text{Water (lb)} + \text{Chemical (lb)}} \times 100 \tag{3.9}$$

### ■ Example 3.11

*Problem:* If 10 ounces (oz.) of dry polymer are added to 15 gal of water, what is the percent strength (by weight) of the polymer solution?

*Solution:* Before calculating percent strength, the ounces of chemical must be converted to pounds of chemical:

$$10 \text{ oz.} \div 16 \text{ oz./lb} = 0.625 \text{ lb chemical}$$

Now calculate percent strength:

$$\text{Percent strength} = \frac{\text{Chemical (lb)}}{\text{Water (lb)} + \text{Chemical (lb)}} \times 100$$

$$= \frac{0.625 \text{ lb}}{(15 \text{ gal} \times 8.34 \text{ lb/gal}) + 0.625 \text{ lb}} \times 100 = 0.5\%$$

### ■ Example 3.12

*Problem:* If 90 g (1 g = 0.0022 lb) of dry polymer are dissolved in 6 gal of water, what percent strength is the solution?

*Solution:* First, convert grams chemical to pounds chemical. Because 1 g equals 0.0022 lb, 90 g is 90 times 0.0022 lb:

$$90 \text{ g polymer} \times 0.0022 \text{ lb/g} = 0.198 \text{ lb}$$

Now calculate percent strength of the solution:

$$\text{Percent strength} = \frac{\text{Chemical (lb)}}{\text{Water (lb)} + \text{Chemical (lb)}} \times 100$$

$$= \frac{0.198 \text{ lb}}{(6 \text{ gal} \times 8.34 \text{ lb/gal}) + 0.198 \text{ lb}} \times 100 = 4\%$$

## Determining Percent Strength of Liquid Solutions

When using liquid chemicals to make up solutions (e.g., liquid polymers), a different calculation is required:

$$\frac{\text{Liquid polymer (lb)} \times \text{Liquid polymer (\% strength)}}{100} = \frac{\text{Polymer solution (lb)} \times \text{Polymer solution (\% strength)}}{100} \tag{3.10}$$

### ■ Example 3.13

*Problem:* A 12% liquid polymer is to be used in making up a polymer solution. How many pounds of liquid polymer should be mixed with water to produce 120 lb of a 0.5% polymer solution?

*Solution:*

$$\frac{\text{Liquid polymer (lb)} \times \text{Liquid polymer (\% strength)}}{100} = \frac{\text{Polymer solution (lb)} \times \text{Polymer solution (\% strength)}}{100}$$

$$\frac{x \text{ lb} \times 12\%}{100} = \frac{120 \text{ lb} \times 0.5\%}{100}$$

$$x \text{ lb} = \frac{120 \times 0.005}{0.12}$$

$$x = 5 \text{ lb}$$

## Determining Percent Strength of Mixed Solutions

The percent strength of solution mixture is determined using the following equation:

$$\% \text{ Strength} = \frac{\left(\frac{\text{Solution 1 (lb)} \times \text{Solution 1 (\% strength)}}{100}\right) + \left(\frac{\text{Solution 2 (lb)} \times \text{Solution 2 (\% strength)}}{100}\right)}{\text{Solution 1 (lb)} + \text{Solution 2 (lb)}} \times 100 \tag{3.11}$$

### ■ Example 3.14

*Problem:* If 12 lb of a 10% strength solution are mixed with 40 lb of 1% strength solution, what is the percent strength of the solution mixture?

*Solution:*

$$\% \text{ Strength} = \frac{\left(\frac{\text{Solution 1 (lb)} \times \text{Solution 1 (\% strength)}}{100}\right) + \left(\frac{\text{Solution 2 (lb)} \times \text{Solution 2 (\% strength)}}{100}\right)}{\text{Solution 1 (lb)} + \text{Solution 2 (lb)}} \times 100$$

$$= \frac{(12 \text{ lb} \times 0.1) + (40 \text{ lb} \times 0.01)}{12 \text{ lb} + 40 \text{ lb}} \times 100 = \frac{1.2 \text{ lb} + 0.4 \text{ lb}}{52 \text{ lb}} \times 100 = 3.1\%$$

## Dry Chemical Feeder Calibration

Occasionally we need to perform a calibration calculation to compare the actual chemical feed rate with the feed rate indicated by the instrumentation. To calculate the actual feed rate for a dry chemical feeder, place a container under the feeder, weigh the container when empty, and then weigh the container again after a specified length of time (e.g., 30 minutes). The actual chemical feed rate can be calculated using the following equation:

$$\text{Feed rate (lb/min)} = \frac{\text{Chemical applied (lb)}}{\text{Length of application (min)}} \tag{3.12}$$

If desired, the chemical feed rate can be converted to lb/day:

$$\text{Feed rate (lb/day)} = \text{Feed rate (lb/min)} \times 1440 \text{ min/day} \tag{3.13}$$

### ■ Example 3.15

*Problem:* Calculate the actual chemical feed rate in lb/day if a container is placed under a chemical feeder and a total of 2 lb is collected during a 30-minute period.

*Solution:* First calculate the lb/min feed rate:

$$\text{Feed rate} = \frac{\text{Chemical applied (lb)}}{\text{Length of application (min)}} = \frac{2 \text{ lb}}{30 \text{ min}} = 0.06 \text{ lb/min}$$

Then calculate the lb/day feed rate using Equation 3.13:

$$\text{Feed rate} = 0.06 \text{ lb/min} \times 1440 \text{ min/day} = 86.4 \text{ lb/day}$$

### ■ Example 3.16

*Problem:* Calculate the actual chemical feed rate in lb/day if a container is placed under a chemical feeder and a total of 1.6 lb is collected during a 20-minute period.

*Solution:* First calculate the lb/min feed rate using Equation 3.12:

$$\text{Feed rate} = \frac{\text{Chemical applied (lb)}}{\text{Length of application (min)}} = \frac{1.6 \text{ lb}}{20 \text{ min}} = 0.08 \text{ lb/min}$$

Then calculate the lb/day feed rate:

$$\text{Feed rate} = 0.08 \text{ lb/min} \times 1440 \text{ min/day} = 115 \text{ lb/day}$$

## Chemical Solution Feeder Calibration

As with other calibration calculations, the actual solution chemical feed rate is determined and then compared with the feed rate indicated by the instrumentation. To calculate the actual solution chemical feed rate, first express the solution feed rate in MGD. Once the MGD solution flow rate has been calculated, use the mg/L equation to determine chemical dosage in lb/day. If solution feed is expressed as mL/min, first convert mL/min flow rate to gpd feed rate.

$$\text{Feed rate (gpd)} = \frac{\text{Feed rate (mL/min)} \times 1440 \text{ min/day}}{3785 \text{ mL/gal}} \quad (3.14)$$

Then calculate chemical dosage in lb/day:

$$\text{Chemical dosage} = \text{Chemical (mg/L)} \times \text{Flow (MGD)} \times 8.34 \text{ lb/day} \quad (3.15)$$

### ■ Example 3.17

*Problem:* A calibration test is conducted for a solution chemical feeder. During a 5-minute test, the pump delivered 940 mg/L of the 1.20% polymer solution. What is the polymer dosage rate in lb/day? (Assume the polymer solution weighs 8.34 lb/gal.)

*Solution:* The feed rate must be expressed as MGD; therefore, the mL/min solution feed rate must first be converted to gpd and then MGD. The mL/min feed rate is calculated as

$$(940 \text{ mL}) \div (5 \text{ min}) = 188 \text{ mL/min}$$

Next convert the mL/min feed rate to gpd feed rate:

$$\text{Feed rate} = \frac{188 \text{ mL/min} \times 1440 \text{ min/day}}{3785 \text{ mL/gal}} = 72 \text{ gpd}$$

Then calculate the lb/day polymer feed rate:

$$\text{Feed rate} = 12{,}000 \text{ mg/L} \times 0.000072 \text{ MGD} \times 8.34 \text{ lb/day} = 7.2 \text{ lb/day}$$

### ■ Example 3.18

*Problem:* A calibration test is conducted for a solution chemical feeder. During a 24-hour period, the solution feeder delivers a total of 100 gal of solution. The polymer solution is a 1.2% solution. What is the lb/day feed rate? (Assume the polymer solution weighs 8.34 lb/gal.)

*Solution:* The solution feed rate is 100 gpd. Expressed as MGD, this is 0.000100 MGD. Use Equation 3.15 to calculate the actual feed rate in lb/day:

$$\text{Feed rate} = \text{Chemical (mg/L)} \times \text{Flow (MGD)} \times 8.34\ \text{lb/day}$$

$$= 12{,}000\ \text{mg/L} \times 0.000100\ \text{MGD} \times 8.34\ \text{lb/day} = 10\ \text{lb/day}$$

The actual pumping rates can be determined by calculating the volume pumped during a specified time frame. For example, if 60 gallons are pumped during a 10-minute test, the average pumping rate during the test is 6 gpm. Actual volume pumped is indicated by a drop in tank level. By using the following equation, we can determine the flow rate in gpm:

$$\text{Flow rate (gpm)} = \frac{0.785 \times D^2 \times \text{Drop in level (ft)} \times 7.48\ \text{gal/ft}^3}{\text{Duration of test (min)}} \tag{3.16}$$

### ■ Example 3.19

*Problem:* A pumping rate calibration test is conducted for a 15-minute period. The liquid level in the 4-ft-diameter solution tank is measured before and after the test. If the level drops 0.5 ft during the 15-minute test, what is the pumping rate in gpm?

*Solution:*

$$\text{Pumping rate} = \frac{0.785 \times D^2 \times \text{Drop in level (ft)} \times 7.48\ \text{gal/ft}^3}{\text{Duration of test (min)}}$$

$$= \frac{0.785 \times (4\ \text{ft})^2 \times 0.5\ \text{ft} \times 7.48\ \text{gal/ft}^3}{15\ \text{min}} = 3.1\ \text{gpm}$$

## Determining Chemical Usage

One of the primary functions performed by water operators is the recording of data. The lb/day or gpd chemical use is part of such data from which the average daily use of chemicals and solutions can be determined. This information is important in forecasting expected chemical use, comparing it with chemicals in inventory, and determining when additional chemicals will be required. To determine average chemical use, we use Equation 3.17 (lb/day) or Equation 3.18 (gpd):

$$\text{Average use (lb/day)} = \frac{\text{Total chemical used (lb)}}{\text{Number of days}} \tag{3.17}$$

$$\text{Average use (gpd)} = \frac{\text{Total chemical used (gal)}}{\text{Number of days}} \tag{3.18}$$

Then we can calculate the days supply in inventory:

$$\text{Days supply in inventory} = \frac{\text{Total chemical in inventory (lb)}}{\text{Average use (lb/day)}} \tag{3.19}$$

$$\text{Days supply in inventory} = \frac{\text{Total chemical in inventory (gal)}}{\text{Average use (gpd)}} \tag{3.20}$$

### ■ Example 3.20

*Problem:* The chemical used each day for a week is given below. Based on the information below, what was the average lb/day chemical use during the week?

Monday, 88 lb/day
Tuesday, 93 lb/day
Wednesday, 91 lb/day
Thursday, 88 lb/day
Friday, 96 lb/day
Saturday, 92 lb/day
Sunday, 86 lb/day

*Solution:*

$$\text{Average use} = \frac{\text{Total chemical used (lb)}}{\text{Number of days}} = \frac{634\text{ lb}}{7\text{ days}} = 90.6\text{ lb/day}$$

### ■ Example 3.21

*Problem:* The average chemical use at a plant is 77 lb/day. If the chemical inventory is 2800 lb, how many days supply is this?

*Solution:*

$$\text{Days supply} = \frac{\text{Total chemical in inventory (lb)}}{\text{Average use (lb/day)}} = \frac{2800\text{ lb}}{77\text{ lb/day}} = 36.4\text{ days}$$

**Note:** Jar tests are performed as required on settling tank influent and are beneficial in determining the best flocculant aid and appropriate doses to improve solids capture during periods of poor settling.

## COAGULATION AND FLOCCULATION PRACTICE PROBLEMS

*Problem 3.1:* The average flow for a water plant is 5.0 MGD. A jar test indicates that the best alum dosage is 3.2 mg/L. How many pounds per day will the operator feed?

*Solution:*

$$\text{Feed rate} = \text{Dose} \times \text{Flow} \times 8.34 \text{ lb/gal}$$

$$= 3.2 \text{ mg/L} \times 5.0 \text{ MGD} \times 8.34 \text{ lb/gal} = 133.4 \text{ lb/day}$$

*Problem 3.2:* The average flow for a waterworks is 1,100,000 gpd. A jar test indicates that the best alum dosage is 1.6 mg/L. How many grams per minute should the feeder deliver?

*Solution:*

$$\text{Feed rate} = \frac{\text{Dose} \times \text{Flow} \times 3.785 \text{ L/gal}}{1440 \text{ min/day} \times 1000 \text{ mg/g}}$$

$$= \frac{1.6 \text{ mg/L} \times 1{,}100{,}000 \text{ gpd} \times 3.785 \text{ L/gal}}{1440 \text{ min/day} \times 1000 \text{ mg/g}} = 4.63 \text{ g/min}$$

*Problem 3.3:* A waterworks used 40 lb of alum to treat 3.5 MGD. Calculate the dose in mg/L.

*Solution:*

$$\text{Dose} = \frac{\text{Feed rate (lb/day)}}{\text{Flow (MGD)} \times 8.34 \text{ lb/gal}} = \frac{40 \text{ lb}}{3.5 \text{ MGD} \times 8.34 \text{ lb/gal}} = 1.37 \text{ mg/L}$$

*Problem 3.4:* Liquid polymer is supplied to a water treatment plant as a 8% solution. How many gallons of this liquid are required to make 140 gal of 1.5% polymer solution?

*Solution:*

$$C_1 \times V_1 = C_2 \times V_2$$

$$0.08 \times V_1 = 0.015 \times 140 \text{ gal}$$

$$V_1 = \frac{0.015 \times 140 \text{ gal}}{0.08}$$

$$= 26.25 \text{ gal}$$

*Problem 3.5:* Liquid alum delivered to a waterworks contains 801.5 mg of alum per milliliter of liquid solution. Jar tests indicate that the best alum dose is 5 mg/L. Determine the setting on the liquid alum feeder in milliliters per minute if that flow is 2.5 MGD.

*Solution:*

$$\text{Feed rate} = \frac{\text{Dose} \times \text{Flow (gpd)} \times 3.785\ \text{L/gal}}{\text{Concentration (mg/mL)} \times 1440\ \text{min/day}}$$

$$= \frac{5\ \text{mg/L} \times 2{,}500{,}000\ \text{gpd} \times 3.785\ \text{L/gal}}{801.5\ \text{mg/mL} \times 1440\ \text{min/day}} = 41\ \text{mL/min}$$

*Problem 3.6:* A waterworks operator switches from dry alum to liquid alum. If the operator feeds an average of 160 lb of dry alum a day, how many gallons of liquid alum must be fed on average given the following information?

Alum liquid, 48% concentration
10.5 lb dry alum per gallon (weight)
5.5 lb dry alum per gallon (concentration)
1.335 specific gravity

*Solution:*

$$\text{Liquid alum} = 160\ \text{lb} \times \frac{1\ \text{gal}}{5.5\ \text{lb alum}} = 29.1\ \text{gal}$$

*Problem 3.7:* The average flow for a waterworks is 3.5 MGD. A jar test indicates that the best alum dosage is 2.2 mg/L. How man pounds per day will the operator feed?

*Solution:*

$$\text{Feed rate} = \text{Dose} \times \text{Flow} \times 8.34\ \text{lb/gal}$$

$$= 2.2\ \text{mg/L} \times 3.5\ \text{MGD} \times 8.34\ \text{lb/gal} = 64.22\ \text{lb/day}$$

*Problem 3.8:* The average flow for a water treatment plant is 12.8 MGD. The jar test indicates that the best alum dose is 1.8 mg/L. How many pounds per day should the operator feed?

*Solution:*

$$\text{Feed rate} = \text{Dose} \times \text{Flow} \times 8.34\ \text{lb/gal}$$

$$= 1.8\ \text{mg/L} \times 12.8\ \text{MGD} \times 8.34\ \text{lb/gal} = 192.2\ \text{lb/day}$$

*Problem 3.9:* Determine the setting on a dry alum feeder in pounds per day when the flow is 1.4 MGD. Jar tests indicate that the best alum dose is 14 mg/L.

*Solution:*

$$\text{Feed rate} = \text{Dose} \times \text{Flow} \times 8.34\ \text{lb/gal}$$

$$= 14\ \text{mg/L} \times 1.4\ \text{MGD} \times 8.34\ \text{lb/gal} = 163.5\ \text{lb/day}$$

*Problem 3.10:* The average flow for a water plant is 8.5 MGD. A jar test indicates that the best alum dosage is 2.4 mg/L. How many grams per minute should the feeder deliver?

*Solution:*

$$\text{Feed rate} = \frac{\text{Dose} \times \text{Flow} \times 3.785 \text{ L/gal}}{1440 \text{ min/day} \times 1000 \text{ mg/g}}$$

$$= \frac{2.4 \text{ mg/L} \times 8{,}500{,}000 \text{ gpd} \times 3.785 \text{ L/gal}}{1440 \text{ min/day} \times 1000 \text{ mg/g}} = 53.6 \text{ g/min}$$

*Problem 3.11:* The average daily flow for a waterworks is 0.85 MGD. If the polymer dosage is kept at 1.9 mg/L, how many pounds of polymer will be used in 30 days?

*Solution:*

$$\text{Polymer} = 1.9 \text{ mg/L} \times 0.85 \times 8.34 \text{ lb/gal} = 13.47 \text{ lb/day}$$

$$13.47 \text{ lb/day} \times 30 \text{ days} = 404.1 \text{ lb}$$

*Problem 3.12:* The average flow for a water plant is 9050 gpm. A jar test indicates that the best polymer dose is 3.2 mg/L. How many pounds will the plant feed in one week? (Assume the plant runs 24 hr/day, 7 days/wk.)

*Solution:*

$$\text{Feed rate} = 9050 \text{ g/min} \times 1440 \text{ min/day} \times 1 \text{ MG/1,000,000 gal} = 13.03 \text{ MGD}$$

$$\text{Feed rate} = 3.2 \text{ m/L} \times 13.03 \text{ MGD} \times 8.34 \text{ lb/gal}$$

$$= 347.74 \text{ lb/day} \times 7 \text{ days/wk}$$

$$= 2434.18 \text{ lb/wk}$$

*Problem 3.13:* A water treatment plant used 30 lb of cationic polymer to treat 1.8 MG of water during a 24-hr period. What is the polymer dosage in mg/L?

*Solution:*

$$\text{Dosage} = \frac{\text{Feed rate (lb/day)}}{\text{Flow (MGD)} \times 8.34 \text{ lb/gal}} = \frac{30 \text{ lb/day}}{1.8 \text{ MGD} \times 8.34 \text{ lb/gal}} = 2.0 \text{ mg/L}$$

*Problem 3.14:* A waterworks fed 140 lb of alum treats 1.5 MGD. Calculate the dose in mg/L.

*Solution:*

$$\text{Dosage} = \frac{\text{Feed rate (lb/day)}}{\text{Flow (MGD)} \times 8.34 \text{ lb/gal}} = \frac{130 \text{ lb/day}}{1.5 \text{ MGD} \times 8.34 \text{ lb/gal}} = 10.4 \text{ mg/L}$$

*Problem 3.15:* A waterworks is fed 50 g/min of dry alum while treating 2.8 MGD. Calculate the mg/L dose.

*Solution:*

$$\text{Flow rate} = 2.8\ \text{MGD} \times (1\ \text{day}/1440\ \text{min}) \times (1{,}000{,}000\ \text{gal}/1\ \text{MG}) = 1944.4\ \text{gpm}$$

$$\text{Dosage} = \frac{\text{Feed rate (g/min)} \times 1000\ \text{mg/g}}{\text{Flow rate (gpm)} \times 3785\ \text{L/gal}} = \frac{50\ \text{g/min} \times 1000\ \text{mg/g}}{1944.4\ \text{gpm} \times 3785\ \text{L/gal}} = 6.8\ \text{mg/L}$$

*Problem 3.16:* Liquid polymer is supplied to a waterworks as an 8% solution. How many gallons of this liquid polymer should be used to make 220 gal of a 0.8% polymer solution?

*Solution:*

$$C_1 \times V_1 = C_2 \times V_2$$

$$0.08 \times V_1 = 0.008 \times 220\ \text{gal}$$

$$V_1 = \frac{0.008 \times 220\ \text{gal}}{0.08} = 22\ \text{gal}$$

*Problem 3.17:* Liquid polymer is supplied to a water treatment plant as a 7% solution. How many gallons of this liquid polymer should be used to make 6 gal of a 6% polymer solution?

*Solution:*

$$C_1 \times V_1 = C_2 \times V_2$$

$$0.07 \times V_1 = 0.06 \times 6\ \text{gal}$$

$$V_1 = \frac{0.06 \times 6\ \text{gal}}{0.07} = 5.14\ \text{gal}$$

*Problem 3.18:* Liquid polymer is supplied to a water treatment plant as a 7% solution. How many gallons of liquid polymer should be used to make 60 gal of a 0.5% polymer solution?

*Solution:*

$$C_1 \times V_1 = C_2 \times V_2$$

$$0.07 \times V_1 = 0.005 \times 60\ \text{gal}$$

$$V_1 = \frac{0.005 \times 60\ \text{gal}}{0.07} = 4.29\ \text{gal}$$

*Problem 3.19:* Liquid alum delivered to a waterworks contains 653.4 mg of alum per milliliter of liquid solution. Jar tests indicate that the best alum dose is 9 mg/L. Determine the setting on the liquid alum chemical feed in milliliters per minute if the flow is 2.3 MGD.

*Solution:*

$$\text{Feed rate} = \frac{\text{Dose (mg/L)} \times \text{Flow (gpd)} \times 3.785\ \text{L/gal}}{\text{Concentration (mg/mL)} \times 1440\ \text{min/day}}$$

$$= \frac{9\ \text{mg/L} \times 2{,}300{,}000\ \text{gpd} \times 3.785\ \text{L/gal}}{653.4\ \text{mg/mL} \times 1440\ \text{min/day}} = 83.3\ \text{mL/min}$$

*Problem 3.20:* Three 2-minute samples are collected from an alum dry feeder. What is the feed rate in mg/L when the flow rate is 3 MGD?

Sample 1 = 22 g
Sample 2 = 25 g
Sample 3 = 23 g

*Solution:*

$$\text{Average} = \frac{22\ \text{g} + 25\ \text{g} + 23\ \text{g}}{3} = 23.3\ \text{g}$$

$$\text{Feed rate} = 23.3\ \text{g/2 min} = 11.7\ \text{g/min}$$

$$\text{Flow rate} = (3{,}000{,}000\ \text{gpd}) \times (1\ \text{day/1440 min}) = 2083.3\ \text{gpm}$$

$$\text{Feed rate} = \frac{\text{Feed rate} \times 1000\ \text{mg/g}}{\text{Flow rate} \times 3.785\ \text{L/gal}} = \frac{(11.7\ \text{g/min}) \times (1000\ \text{mg/g})}{2083.3\ \text{gpm} \times 3.785\ \text{L/gal}} = 1.48\ \text{mg/L}$$

*Problem 3.21:* A waterworks is treating 9.1 MGD with 2.0 mg/L liquid alum. How many gpd of liquid alum will be required? The liquid alum contains 5.25 lb dry alum per gallon.

*Solution:*

$$\text{Liquid alum} = \frac{\text{Dose (mg/L)} \times \text{Flow (MGD)} \times 8.34\ \text{lb/gal}}{\text{Concentration (lb/gal)}}$$

$$= \frac{2.0\ \text{mg/L} \times 9.1\ \text{MGD} \times 8.34\ \text{lb/gal}}{5.25\ \text{lb/gal}} = 28.91\ \text{gpd}$$

*Problem 3.22:* A jar test indicates that 3.2 mg/L of liquid alum are required to treat 7.5 MGD. How many mL/min should the metering pump deliver? The liquid alum delivered to the plant contains 640 mg alum per mL of liquid solution.

*Solution:*

$$\text{Feed rate} = \frac{3.2\ \text{mg/L} \times 7{,}500{,}000\ \text{gpd} \times 3.785\ \text{L/gal}}{640\ \text{mg/mL} \times 1440\ \text{min/day}} = 98.57\ \text{mL/min}$$

*Problem 3.23:* A jar test indicates that 1.9 mg/L of liquid ferric chloride should be fed to treat 2878 gpm of water. How many mL/min should be fed by a metering pump? Ferric chloride contains 4.6 lb dry chemical per gallon of liquid solution.

*Solution:*

$$\text{Flow rate} = 2878\ \text{gpm} \times 1440\ \text{min/day} = 4{,}144{,}320\ \text{gpd}$$

$$\text{Feed rate (mg/mL)} = (4.6\ \text{lb/gal}) \times (1\ \text{gal/3785 mL}) \times (453.6\ \text{g/1 lb}) \times 1000\ (\text{mg/1 g})$$

$$= 551.27\ \text{mg/mL}$$

$$\text{Feed rate (mL/min)} = \frac{1.9\ \text{mg/L} \times 4{,}144{,}320\ \text{gpd} \times 3.785\ \text{mL/gal}}{551.27\ \text{mg/mL} \times 1440\ \text{min/day}} = 37.54\ \text{mL/min}$$

## CHEMICAL FEEDERS

Simply put, a chemical feeder is a mechanical device for measuring a quantity of chemical and applying it to water at a preset rate. Two types of chemical feeders are commonly used: solution (or liquid) feeders and dry feeders. Liquid feeders apply chemicals in solutions or suspensions, and dry feeders apply chemicals in granular or powdered forms. In a solution feeder, the chemical enters and leaves the feeder in a liquid state; in a dry feeder, the chemical enters and leaves the feeder in a dry state.

### SOLUTION FEEDERS

Solution feeders are small, positive-displacement metering pumps of three types: (1) reciprocating (piston-plunger or diaphragm types), (2) vacuum (e.g., gas chlorinator), or (3) gravity-feed rotameter (e.g., drip feeder). Positive-displacement pumps are used in high-pressure, low-flow applications; they deliver a specific volume of liquid for each stroke of a piston or rotation of an impeller.

### DRY FEEDERS

Two types of dry feeders are volumetric and gravimetric, depending on whether the chemical is measured by volume (volumetric type) or weight (gravimetric type). Simpler and less expensive than gravimetric pumps, volumetric dry feeders are also less accurate. Gravimetric dry feeders are extremely accurate, deliver high feed rates, and are more expensive than volumetric feeders.

### Chemical Feeder Calibration

Chemical feeder calibration ensures effective control of the treatment process. Obviously, chemical feed without some type of metering and accounting of chemical used adversely affects the water treatment process. Chemical feeder calibration also optimizes economy of operation; it ensures the optimum use of expensive chemicals. Finally, operators must have accurate knowledge of each individual feeder's capabilities at specific settings. When a certain dose must be administered, the operator must rely on the feeder to feed the correct amount of chemical. Proper calibration ensures that chemical dosages can be set with confidence. At a minimum, chemical feeders must be calibrated on an annual basis. During operation, when the operator changes chemical strength or chemical purity or makes any adjustment to the feeder, or when the treated water flow changes, the chemical feeder should be calibrated. Ideally, any time maintenance is performed on chemical feed equipment, calibration should be performed.

What factors affect chemical feeder calibration (i.e., feed rate)? For solution feeders, calibration is affected any time solution strength changes, any time a mechanical change is introduced in the pump (change in stroke length or stroke frequency), and whenever the flow rate changes. In the dry chemical feeder, calibration is affected any time the chemical purity changes, mechanical damage occurs (e.g., belt change), or the flow rate changes. In the calibration process, calibration charts are usually used or made up to fit the calibration equipment. The calibration chart is affected by certain factors, including change in chemical, change in flow rate of water being treated, and a mechanical change in the feeder.

> ***Note:*** Pounds per day (lb/day) is not normally useful information for setting the feed rate setting on a feeder, because process control usually determines a dosage in ppm, mg/L, or grains/gal. A separate chart may be necessary for another conversion based on the individual treatment facility flow rate.

## CHEMICAL FEEDER CALCULATIONS

### ■ Example 3.22

*Problem:* An operator collects three 3-minute samples from a dry feeder. Based on the information given below, determine the average gpm.

Sample 1 = 36.8 g
Sample 2 = 37.8 g
Sample 3 = 35.9 g

*Solution:*

$$\text{Average} = \frac{36.8\text{ g} + 37.8\text{ g} + 35.9\text{ g}}{3} = 36.8\text{ g}$$

$$\text{Flow rate} = \frac{36.8\text{ g/min}}{3\text{ min}} = 12.28\text{ g/min}$$

■ **EXAMPLE 3.23**

*Problem:* What is the average dose in mg/L for the feeder in the previous question if the plant treats 3.7 MGD?

*Solution:*

$$3{,}700{,}000 \text{ gal/day} \times 1 \text{ day/60 min} = 61{,}667 \text{ gpm}$$

$$\text{Dosage} = \frac{\text{Feed rate (g/min)} \times 1000 \text{ mg/g}}{\text{Flow rate (gpm)} \times 3.785 \text{ L/gal}}$$

$$= \frac{12.28 \text{ g/min} \times 1000 \text{ mg/g}}{61{,}667 \text{ gpm} \times 3.785 \text{ L/gal}} = 0.053 \text{ mg/L}$$

■ **EXAMPLE 3.24**

*Problem:* An operator is checking the calibration on a chemical feeder. The feeder delivers 105 g in 5 minutes. How many grams per minutes does the feeder deliver? How many pounds per day does the feeder deliver?

*Solution:*

$$\text{Feed rate (g/min)} = \frac{105 \text{ g}}{5 \text{ min}} = 21 \text{ g/min}$$

$$\text{Feed rate (lb/day)} = 21 \text{ g/min} \times 1440 \text{ min/day} \times 1 \text{ lb/453.6 g} = 66.7 \text{ lb/day}$$

■ **EXAMPLE 3.25**

*Problem:* An operator checks the calibration of a dry feeder by catching samples and weighing them on a balance. Each catch lasts 1 minute. Calculate the average feed rate in grams per minute based on the following data and determine how many pounds per hour are being fed.

Sample 1 = 38.0 g
Sample 2 = 36.3 g
Sample 3 = 39.2 g
Sample 4 = 38.5 g

*Solution:*

$$\text{Average} = \frac{38.0 \text{ g} + 36.3 \text{ g} + 39.2 \text{ g} + 38.5 \text{ g}}{4} = 38 \text{ g}$$

$$\text{Flow rate} = 38 \text{ g/1 min} = 38 \text{ g/min}$$

$$\text{Feed rate (mg/L)} = (38 \text{ g/min}) \times (60 \text{ min/hr}) \times (1 \text{ lb/453.6 g}) = 5.03 \text{ lb/hr}$$

### ■ EXAMPLE 3.26

*Problem:* An operator collects three 2-minute samples from a dry feeder:

Sample 1 = 21.3 g
Sample 2 = 24.2 g
Sample 3 = 21.7 g

What is the average grams per minute? What is the average dose in mg/L for the feeder if the plant treats 430,000 gpd?

*Solution:*

$$\text{Average} = \frac{21.3\text{ g} + 24.2\text{ g} + 21.7\text{ g}}{3} = 22.4\text{ g}$$

$$\text{Feed rate} = 22.4\text{ g/2 min} = 11.2\text{ g/min}$$

$$\text{Flow rate} = (4{,}300{,}000\text{ gpd}) \times (1\text{ day/1440 min}) = 2986.1\text{ gpm}$$

$$\text{Dosage} = \frac{\text{Feed rate (g/min)} \times 1000\text{ mg/g}}{\text{Flow rate (gpm)} \times 3.785\text{ L/gal}} = \frac{11.2\text{ g/min} \times 1000\text{ mg/g}}{2986.1\text{ gpm} \times 3.785\text{ L/gal}} = 0.99\text{ mg/L}$$

### ■ EXAMPLE 3.27

*Problem:* An operator collects five 2-minute samples from a dry feeder:

Sample 1 = 49.4 g
Sample 2 = 44.2 g
Sample 3 = 41.8 g
Sample 4 = 48.4 g
Sample 5 = 47.9 g

What is the average number of grams per minute? What is the average dose in mg/L if the plant treats 1,300,000 gpd?

*Solution:*

$$\text{Average} = \frac{49.4\text{ g} + 44.2\text{ g} + 41.8\text{ g} + 48.4\text{ g} + 47.9\text{ g}}{5} = 46.34\text{ g}$$

$$\text{Feed rate} = 46.34\text{ g/2 min} = 23.17\text{ g/min}$$

$$\text{Flow rate} = (1{,}300{,}000\text{ gpd}) \times (1\text{ day/1440 min}) = 902.8\text{ gpm}$$

$$\text{Dosage} = \frac{\text{Feed rate (g/min)} \times 1000\text{ mg/g}}{\text{Flow rate (gpm)} \times 3.785\text{ L/gal}} = \frac{23.17\text{ g/min} \times 1000\text{ mg/g}}{902.8\text{ gpm} \times 3.785\text{ L/gal}} = 6.78\text{ mg/L}$$

### ■ EXAMPLE 3.28

*Problem:* A chemical feeder calibration is tested using a 1000-mL graduated cylinder. The cylinder is filled to 830 mL in a 3-minute test. What is the chemical feed rate in milliliters per minute? What is the chemical feed rate in gallons per minute? What is the chemical feed rate in gallons per day?

*Solution:*

$$\text{Feed rate (mL/min)} = 830 \text{ mL}/3 \text{ min} = 276.7 \text{ mL/min}$$

$$\text{Feed rate (gpm)} = (276.7 \text{ mL/min}) \times (1 \text{ L}/1000 \text{ mL}) \times (1 \text{ gal}/3.785 \text{ L}) = 0.071 \text{ gpm}$$

$$\text{Feed rate (gpd)} = (0.071 \text{ gal/min}) \times (1440 \text{ min}/1 \text{ day}) = 102.24 \text{ gpd}$$

# 4 Sedimentation Calculations

## SEDIMENTATION

Sedimentation—solid–liquid separation by gravity—is one of the most basic processes of water and wastewater treatment. In water treatment, plain sedimentation, such as the use of a pre-sedimentation basin for grit removal and a sedimentation basin following coagulation–flocculation, is the most commonly used type.

## TANK VOLUME CALCULATIONS

Sedimentation tanks are generally rectangular or cylindrical. The equations for calculating the volume for each type of tank are shown below.

### Calculating Tank Volume

For rectangular sedimentation basins, we use Equation 4.1:

$$\text{Volume (gal)} = \text{Length (ft)} \times \text{Width (ft)} \times \text{Depth (ft)} \times 7.48\ \text{gal/ft}^3 \tag{4.1}$$

For circular clarifiers, we use Equation 4.2:

$$\text{Volume (gal)} = 0.785 \times D^2 \times \text{Depth (ft)} \times 7.48\ \text{gal/ft}^3 \tag{4.2}$$

■ **Example 4.1**

*Problem:* A sedimentation basin is 25 ft wide by 80 ft long and contains water to a depth of 14 ft. What is the volume of water in the basin in gallons?

*Solution:*

$$\begin{aligned}\text{Volume} &= \text{Length (ft)} \times \text{Width (ft)} \times \text{Depth (ft)} \times 7.48\ \text{gal/ft}^3\\ &= 80\ \text{ft} \times 25\ \text{ft} \times 14\ \text{ft} \times 7.48\ \text{gal/ft}^3 = 209{,}440\ \text{gal}\end{aligned}$$

■ **Example 4.2**

*Problem:* A sedimentation basin is 24 ft wide by 75 ft long. When the basin contains 140,000 gallons, what would the water depth be?

*Solution:*

$$\text{Volume} = \text{Length (ft)} \times \text{Width (ft)} \times \text{Depth (ft)} \times 7.48\ \text{gal/ft}^3$$

$$140{,}000\ \text{gal} = 75\ \text{ft} \times 24\ \text{ft} \times x\ \text{ft} \times 7.48\ \text{gal/ft}^3$$

$$x\ \text{ft} = \frac{140{,}000\ \text{gal}}{75\ \text{ft} \times 24\ \text{ft} \times 7.48\ \text{gal/ft}^3}$$

$$x = 10.4\ \text{ft}$$

## DETENTION TIME

Detention time for clarifiers varies from 1 to 3 hours. The equations used to calculate detention time are shown below.

- Basic detention time equation

$$\text{Detention time (hr)} = \frac{\text{Volume of tank (gal)}}{\text{Flow rate (gph)}} \tag{4.3}$$

- Rectangular sedimentation basin equation

$$\text{Detention time (hr)} = \frac{\text{Length (ft)} \times \text{Width (ft)} \times \text{Depth (ft)} \times 7.48\ \text{gal/ft}^3}{\text{Flow rate (gph)}} \tag{4.4}$$

- Circular basin equation

$$\text{Detention time (hr)} = \frac{0.785 \times D^2 \times \text{Depth (ft)} \times 7.48\ \text{gal/ft}^3}{\text{Flow rate (gph)}} \tag{4.5}$$

### ■ EXAMPLE 4.3

*Problem:* A sedimentation tank has a volume of 137,000 gal. If the flow to the tank is 121,000 gallons per hour (gph), what is the detention time in the tank in hours?

*Solution:*

$$\text{Detention time} = \frac{\text{Volume of tank (gal)}}{\text{Flow rate (gph)}} = \frac{137{,}000\ \text{gal}}{121{,}000\ \text{gph}} = 1.1\ \text{hr}$$

### ■ EXAMPLE 4.4

*Problem:* A sedimentation basin is 60 ft long by 22 ft wide and has water to a depth of 10 ft. If the flow to the basin is 1,500,000 gallons per day (gpd), what is the sedimentation basin detention time?

*Solution:* First, convert the flow rate from gpd to gph so the times units will match (1,500,000 gpd ÷ 24 hr/day = 62,500 gph). Then calculate detention time:

$$\text{Detention time} = \frac{\text{Length (ft)} \times \text{Width (ft)} \times \text{Depth (ft)} \times 7.48 \text{ gal/ft}^3}{\text{Flow rate (gph)}}$$

$$= \frac{60 \text{ ft} \times 22 \text{ ft} \times 10 \text{ ft} \times 7.48 \text{ gal/ft}^3}{62{,}500 \text{ gph})}$$

$$= 1.6 \text{ hr}$$

## SURFACE LOADING (OVERFLOW) RATE

Surface loading (overflow) rate—similar to hydraulic loading rate (flow per unit area)—is used to determine loading on sedimentation basins and circular clarifiers. Hydraulic loading rate, however, measures the total water entering the process, whereas surface loading rate measures only the water overflowing the process (plant flow only).

> ***Note:*** Surface loading rate calculations do not include recirculated flows. Other terms used synonymously with surface loading rate are *surface overflow rate* and *surface settling rate.*

Surface loading rate is determined using the following equation:

$$\text{Surface loading rate} = \frac{\text{Flow (gpm)}}{\text{Area (ft}^2\text{)}} \tag{4.6}$$

### ■ Example 4.5

*Problem:* A circular clarifier has a diameter of 80 ft. If the flow to the clarifier is 1800 gallons per minute (gpm), what is the surface loading rate in gpm/ft$^2$?

*Solution:*

$$\text{Surface loading rate} = \frac{\text{Flow (gpm)}}{\text{Area (ft}^2\text{)}} = \frac{1800 \text{ gpm}}{0.785 \times 80 \text{ ft} \times 80 \text{ ft}} = 0.36 \text{ gpm/ft}^2$$

### ■ Example 4.6

*Problem:* A sedimentation basin 70 ft by 25 ft receives a flow of 1000 gpm. What is the surface loading rate in gpm/ft$^2$?

*Solution:*

$$\text{Surface loading rate} = \frac{\text{Flow (gpm)}}{\text{Area (ft}^2\text{)}} = \frac{1000 \text{ gpm}}{70 \text{ ft} \times 25 \text{ ft}} = 0.6 \text{ gpm/ft}^2$$

## MEAN FLOW VELOCITY

The measure of average velocity of the water as it travels through a rectangular sedimentation basin is known as mean flow velocity. Mean flow velocity is calculated as follows:

$$Q \text{ (flow) (ft}^3\text{/min)} = A \text{ (cross-sectional area) (ft}^2) \times V \text{ (volume) (ft/min)} \quad (4.7)$$

$$Q = A \times V$$

### ■ Example 4.7

*Problem:* A sedimentation basin is 60 ft long by 18 ft wide and has water to a depth of 12 ft. When the flow through the basin is 900,000 gpd, what is the mean flow velocity in the basin in ft/min?

*Solution:* Because velocity is desired in ft/min, the flow rate in the $Q = A \times V$ equation must be expressed in ft³/min (cfm):

$$\text{Flow rate} = \frac{900{,}000 \text{ gpd}}{1440 \text{ min/day} \times 7.48 \text{ gal/ft}^3} = 84 \text{ cfm}$$

Then use the $Q = A \times V$ equation to calculate velocity:

$$Q = A \times V$$

$$84 \text{ cfm} = 18 \text{ ft} \times 12 \text{ ft} \times x \text{ fpm}$$

$$x \text{ fpm} = \frac{84 \text{ cfm}}{18 \text{ ft} \times 12 \text{ ft}}$$

$$x = 0.4 \text{ fpm}$$

### ■ Example 4.8

*Problem:* A rectangular sedimentation basin 50 ft long by 20 ft wide has a water depth of 9 ft. If the flow to the basin is 1,880,000 gpd, what is the mean flow velocity in ft/min?

*Solution:* Because velocity is desired in ft/min, the flow rate in the $Q = A \times V$ equation must be expressed in ft³/min (cfm):

$$\text{Flow rate} = \frac{1{,}880{,}000 \text{ gpd}}{1440 \text{ min/day} \times 7.48 \text{ gal/ft}^3} = 175 \text{ cfm}$$

Then use the $Q = A \times V$ equation to calculate velocity:

$$Q = A \times V$$

$$175 \text{ cfm} = 20 \text{ ft} \times 9 \text{ ft} \times x \text{ fpm}$$

$$x = \frac{175 \text{ cfm}}{20 \text{ ft} \times 9 \text{ ft}} = 0.97 \text{ fpm}$$

## WEIR LOADING (OVERFLOW) RATE

The weir loading (overflow) rate is the amount of water leaving the settling tank per linear foot of weir. The result of this calculation can be compared with design. Weir loading rates of 10,000 to 20,000 gal/day/ft are usually used in the design of a settling tank. Typically, the weir loading rate is a measure of the gallons per minute (gpm) flow over each foot of weir. Weir loading rate is determined using the following equation:

$$\text{Weir loading rate (gpm/ft)} = \frac{\text{Flow (gpm)}}{\text{Weir length (ft)}} \tag{4.8}$$

### ■ Example 4.9

*Problem:* A rectangular sedimentation basin has a total of 115 ft of weir. What is the weir loading rate, in gpm/ft, for a flow of 1,110,000 gpd?

*Solution:*

$$\text{Flow} = \frac{1{,}110{,}000 \text{ gpd}}{1440 \text{ min/day}} = 771 \text{ gpm}$$

$$\text{Weir loading rate} = \frac{\text{Flow (gpm)}}{\text{Weir length (ft)}} = \frac{771 \text{ gpm}}{115 \text{ ft}} = 6.7 \text{ gpm/ft}$$

### ■ Example 4.10

*Problem:* A circular clarifier receives a flow of 3.55 MGD. If the diameter of the weir is 90 ft, what is the weir loading rate in gpm/ft?

*Solution:*

$$\text{Flow} = \frac{3{,}550{,}000 \text{ gpd}}{1440 \text{ min/day}} = 2465 \text{ gpm}$$

$$\text{Feet of weir} = 3.14 \times 90 \text{ ft} = 283 \text{ ft}$$

$$\text{Weir loading rate} = \frac{\text{Flow (gpm)}}{\text{Weir length (ft)}} = \frac{2465 \text{ gpm}}{283 \text{ ft}} = 8.7 \text{ gpm/ft}$$

## PERCENT SETTLED BIOSOLIDS

The percent settled biosolids test, also known as the volume over volume (V/V) test, is conducted by collecting a 100-mL slurry sample from the solids contact unit and then allowing the sample to settle for 10 minutes. After 10 minutes, the volume of settled biosolids at the bottom of the 100-mL graduated cylinder is measured and recorded. The equation used to calculate percent settled biosolids is shown below:

$$\%\text{ Settled biosolids} = \frac{\text{Settled biosolids volume (mL)}}{\text{Total sample volume (mL)}} \times 100 \tag{4.9}$$

### ■ Example 4.11

*Problem:* A 100-mL sample of slurry from a solids contact unit is placed in a graduated cylinder and allowed to set for 10 minutes. The volume of settled biosolids at the bottom of the graduated cylinder after 10 minutes is 22 mL. What is the percent of settled biosolids of the sample?

*Solution:*

$$\%\text{ Settled biosolids} = \frac{\text{Settled biosolids volume (mL)}}{\text{Total sample volume (mL)}} \times 100 = \frac{22\text{ mL}}{100\text{ mL}} \times 100 = 19\%$$

### ■ Example 4.12

*Problem:* Suppose that a 100-mL sample of slurry from a solids contact unit is placed in a graduated cylinder. After 10 minutes, a total of 21 mL of biosolids settled to the bottom of the cylinder. What is the percent settled biosolids of the sample?

*Solution:*

$$\%\text{ Settled biosolids} = \frac{\text{Settled biosolids volume (mL)}}{\text{Total sample volume (mL)}} \times 100 = \frac{21\text{ mL}}{100\text{ mL}} \times 100 = 21\%$$

## DETERMINING LIME DOSAGE (MG/L)

During the alum dosage process, lime is sometimes added to provide adequate alkalinity (as bicarbonate, $HCO_3^-$) in the solids contact clarification process for coagulation and precipitation of the solids. To determine the lime dose required, in mg/L, three steps are required.

### Step 1

In Step 1, the total alkalinity required is calculated. The total alkalinity required to react with the alum to be added and to provide proper precipitation is determined using the following equation:

$$\text{Total alk. required (mg/L)} = \text{Alk. reacting with alum (mg/L)} + \text{Alk. in water (mg/L)} \tag{4.10}$$

$$\uparrow$$

(1 mg/L reacts with 0.45 mg/L alkalinity)

### ■ Example 4.13

*Problem:* A raw water requires an alum dose of 45 mg/L, as determined by jar testing. If a residual 30 mg/L alkalinity must be present in the water to ensure complete precipitation of alum added, what is the total alkalinity required in mg/L?

*Solution:* First, calculate the alkalinity that will react with 45 mg/L alum:

$$\frac{0.45 \text{ mg/L alk.}}{1 \text{ mg/L alum}} = \frac{x \text{ mg/L alk.}}{45 \text{ mg/L alum}}$$

$$0.45 \text{ mg/L alk.} \times 45 \text{ mg/L alum} = x \text{ mg/L alk.} \times 1 \text{ mg/L alum}$$

$$20.25 \text{ mg/L alk.} = x$$

Next, calculate the total alkalinity required:

$$\text{Total alk. required} = \text{Alk. reacting with alum (mg/L)} + \text{Alk. in water (mg/L)}$$

$$= 20.25 \text{ mg/L} + 30 \text{ mg/L} = 50.25 \text{ mg/L}$$

### ■ Example 4.14

*Problem:* Jar tests indicate that 36 mg/L alum are optimum for particular raw water. If a residual 30 mg/L alkalinity must be present to promote complete precipitation of the alum added, what is the total alkalinity required in mg/L?

*Solution:* First, calculate the alkalinity that will react with 36 mg/L alum:

$$\frac{0.45 \text{ mg/L alk.}}{1 \text{ mg/L alum}} = \frac{x \text{ mg/L alk.}}{36 \text{ mg/L alum}}$$

$$0.45 \text{ mg/L alk.} \times 36 \text{ mg/L alum} = x \text{ mg/L alk.} \times 1 \text{ mg/L alum}$$

$$16.2 \text{ mg/L alk.} = x$$

Then, calculate the total alkalinity required:

$$\text{Total alk. required} = \text{Alk. reacting with alum (mg/L)} + \text{Alk. in water (mg/L)}$$
$$= 16.2 \text{ mg/L} + 30 \text{ mg/L} = 46.2 \text{ mg/L}$$

### Step 2

In Step 2, we make a comparison between the required alkalinity and the alkalinity already in the raw water to determine how many mg/L alkalinity should be added to the water. The equation used to make this calculation is shown below:

$$\text{Alkalinity to be added to water (mg/L)} = \text{Total alkalinity required (mg/L)} - \text{Alkalinity in water (mg/L)} \quad (4.11)$$

### ■ Example 4.15

*Problem:* A total of 44 mg/L alkalinity is required to react with alum and ensure proper precipitation. If the raw water has an alkalinity of 30 mg/L as bicarbonate, how many mg/L alkalinity should be added to the water?

*Solution:*

$$\text{Alk. to be added to water} = \text{Total alk. required (mg/L)} - \text{Alk. in water (mg/L)}$$
$$= 44 \text{ mg/L} - 30 \text{ mg/L} = 14 \text{ mg/L}$$

### Step 3

In Step 3, after determining the amount of alkalinity to be added to the water, we determine how much lime (the source of alkalinity) needs to be added. We accomplish this by using the ratio shown in Example 4.16.

### ■ Example 4.16

*Problem:* It has been calculated that 16 mg/L alkalinity must be added to a raw water. How many mg/L lime will be required to provide this amount of alkalinity? (One mg/L alum reacts with 0.45 mg/L, and 1 mg/L alum reacts with 0.35 mg/L lime.)

*Solution:* First, determine the mg/L lime required by using a proportion that relates bicarbonate alkalinity to lime:

$$\frac{0.45 \text{ mg/L alk.}}{0.35 \text{ mg/L lime}} = \frac{16 \text{ mg/L alk.}}{x \text{ mg/L lime}}$$

$$0.45 \text{ mg/L alk.} \times x \text{ mg/L lime} = 0.35 \text{ mg/L lime} \times 16 \text{ mg/L alk.}$$

$$x = \frac{0.35 \text{ mg/L lime} \times 16 \text{ mg/L alk.}}{0.45 \text{ mg/L alk.}} = 12.4 \text{ mg/L}$$

In Example 4.17, we use all three steps to determine the lime dosage (mg/L) required.

### ■ Example 4.17

*Problem:* Given the following data, calculate the lime dose required in mg/L:

Alum dose required (determined by jar tests) = 52 mg/L
Residual alkalinity required for precipitation = 30 mg/L
1 mg/L alum reacts with 0.35 mg/L lime
1 mg/L alum reacts with 0.45 mg/L alkalinity
Raw water alkalinity = 36 mg/L

*Solution:* To calculate the total alkalinity required, we must first calculate the alkalinity that will react with 52 mg/L alum:

$$\frac{0.45 \text{ mg/L alk.}}{1 \text{ mg/L alum}} = \frac{x \text{ mg/L alk.}}{36 \text{ mg/L alum}}$$

$$0.45 \text{ mg/L alk.} \times 52 \text{ mg/L alum} = x \text{ mg/L alk.} \times 1 \text{ mg/L alum}$$

$$23.4 \text{ mg/L alk.} = x$$

The total alkalinity requirement can now be determined:

$$\text{Total alk. required} = \text{Alk. reacting with alum (mg/L)} + \text{Alk. in water (mg/L)}$$

$$= 23.4 \text{ mg/L} + 30 \text{ mg/L}$$

$$= 53.4 \text{ mg/L}$$

Next, calculate how much alkalinity must be added to the water:

$$\text{Alk. to be added to water} = \text{Total alk. required (mg/L)} - \text{Alk. in water (mg/L)}$$

$$= 53.4 \text{ mg/L} - 36 \text{ mg/L}$$

$$= 17.4 \text{ mg/L}$$

Finally, calculate the lime required to provide this additional alkalinity:

$$\frac{0.45 \text{ mg/L alk.}}{0.35 \text{ mg/L lime}} = \frac{17.4 \text{ mg/L alk.}}{x \text{ mg/L lime}}$$

$$0.45 \text{ mg/L alk.} \times x \text{ mg/L lime} = 0.35 \text{ mg/L lime} \times 17.4 \text{ mg/L alk.}$$

$$x = \frac{0.35 \text{ mg/L lime} \times 17.4 \text{ mg/L alk.}}{0.45 \text{ mg/L alk.}} = 13.5 \text{ mg/L}$$

## DETERMINING LIME DOSAGE (LB/DAY)

After the lime dose has been determined in terms of mg/L, it is a fairly simple matter to calculate the lime dose in lb/day, which is one of the most common calculations in water and wastewater treatment. To convert from mg/L to lb/day lime dose, we use the following equation:

$$\text{Lime (lb/day)} = \text{Lime (mg/L)} \times \text{Flow (MGD)} \times 8.34\ \text{lb/gal} \tag{4.12}$$

### ■ EXAMPLE 4.18

*Problem:* The lime dose for a raw water has been calculated to be 15.2 mg/L. If the flow to be treated is 2.4 million gallons per day (MGD), how many lb/day lime will be required?

*Solution:*

$$\text{Lime (lb/day)} = \text{Lime (mg/L)} \times \text{Flow (MGD)} \times 8.34\ \text{lb/gal}$$

$$= 15.2\ \text{mg/L} \times 2.4\ \text{MGD} \times 8.34\ \text{lb/gal} = 304\ \text{lb/day}$$

### ■ EXAMPLE 4.19

*Problem:* The flow to a solids contact clarifier is 2,650,000 gpd. If the lime dose required is determined to be 12.6 mg/L, how many lb/day lime will be required?

*Solution:*

$$\text{Lime (lb/day)} = \text{Lime (mg/L)} \times \text{Flow (MGD)} \times 8.34\ \text{lb/gal}$$

$$= 12.6\ \text{mg/L} \times 2.65\ \text{MGD} \times 8.34\ \text{lb/gal} = 278\ \text{lb/day}$$

## DETERMINING LIME DOSAGE (G/MIN)

To convert from mg/L lime to g/min lime, use Equation 4.13.

**Note:** 1 lb = 453.6 g.

$$\text{Lime (g/min)} = \frac{\text{Lime (lb/day)} \times 453.6\ \text{g/lb}}{1440\ \text{min/day}} \tag{4.13}$$

### ■ EXAMPLE 4.20

*Problem:* A total of 275 lb/day lime will be required to raise the alkalinity of the water passing through a solids-contact clarification process. How many g/min lime does this represent?

*Solution:*

$$\text{Lime (g/min)} = \frac{\text{Lime (lb/day)} \times 453.6\ \text{g/lb}}{1440\ \text{min/day}} = \frac{275\ \text{lb/day} \times 453.6\ \text{g/lb}}{1440\ \text{min/day}} = 86.6\ \text{g/min}$$

■ **Example 4.21**

*Problem:* A lime dose of 150 lb/day is required for a solids contact clarification process. How many g/min lime does this represent?

*Solution:*

$$\text{Lime (g/min)} = \frac{\text{Lime (lb/day)} \times 453.6 \text{ g/lb}}{1440 \text{ min/day}} = \frac{150 \text{ lb/day} \times 453.6 \text{ g/lb}}{1440 \text{ min/day}} = 47.3 \text{ g/min}$$

## SEDIMENTATION PRACTICE PROBLEMS

*Problem 4.1:* The flow to a sedimentation tanks is 220,000 gpd. If the tank is 52 ft long by 34 ft wide, what is the surface loading rate in gpd/ft$^2$?

*Solution:*

$$\text{Surface loading rate} = \frac{\text{Flow (gpd)}}{\text{Area (ft}^2\text{)}} = \frac{220{,}000 \text{ gpd}}{52 \text{ ft} \times 34 \text{ ft}} = 124 \text{ gpd/ft}^2$$

*Problem 4.2:* A tank has a length of 80 ft and is 22 ft wide. What is the weir length around the basin in feet?

*Solution:*

$$\text{Weir length} = (2 \times \text{Length}) + (2 \times \text{Width})$$

$$= (2 \times 80 \text{ ft}) + (2 \times 22 \text{ ft}) = 160 \text{ ft} + 44 \text{ ft} = 204 \text{ ft}$$

*Problem 4.3:* A clarifier has a diameter of 80 ft. What is the length of the weir around the clarifier in feet?

*Solution:*

$$\text{Weir length} = 3.14 \times \text{Diameter} = 3.14 \times 80 \text{ ft} = 251.2 \text{ ft}$$

*Problem 4.4:* The diameter of weir in a circular clarifier is 100 ft. What is the weir overflow rate in gpd/ft if the flow over the weir is 1.80 MGD?

*Solution:*

$$\text{Weir overflow rate} = \frac{\text{Flow (gpd)}}{\text{Weir length (ft)}} = \frac{1{,}800{,}000 \text{ gpd}}{3.14 \times 100 \text{ ft}} = 5732.5 \text{ gpd/ft}$$

*Problem 4.5:* A clarifier is 42 ft long by 30 ft wide and 12 ft deep. If the daily flow is 3.3 MGD, what is the detention time in the basin in minutes?

*Solution:*

$$\text{Volume} = 42\text{ ft} \times 30\text{ ft} \times 12\text{ ft} \times 7.48\text{ gal/ft}^3 = 113{,}098\text{ gal}$$

$$\text{Detention time} = \frac{\text{Volume (gal)} \times 24\text{ hr/day}}{\text{Flow (gpd)}}$$

$$= \frac{113{,}098\text{ gal} \times 24\text{ hr/day}}{3{,}300{,}000\text{ gpd}} = 0.823\text{ hr} \times 60\text{ min/hr} = 49.4\text{ min}$$

*Problem 4.6:* A tank has a length of 110 ft, a width of 22 ft, and a depth of 12 ft. What is the surface area in square feet?

*Solution:*

$$\text{Area} = \text{Length (ft)} \times \text{Width (ft)} = 110\text{ ft} \times 22\text{ ft} = 2420\text{ ft}^2$$

*Problem 4.7:* A clarifier has a diameter of 80 ft and a depth of 10 ft. What is the surface area of the clarifier in square feet?

*Solution:*

$$\text{Area} = 0.785 \times D^2\text{ (ft)} = 0.785 \times (80\text{ ft})^2 = 5024\text{ ft}^2$$

*Problem 4.8:* The flow to a sedimentation tank is 3.15 MGD. If the tank is 80 ft long by 22 ft wide, what is the surface loading rate in gallons per day per square foot?

*Solution:*

$$\text{Surface loading rate} = \frac{\text{Flow (gal/day)}}{\text{Area (ft}^2\text{)}} = \frac{3{,}150{,}000\text{ gpd}}{80\text{ ft} \times 22\text{ ft}} = 1789.8\text{ gpd/ft}^2$$

*Problem 4.9:* The flow to a sedimentation tank is 55,000 gpd. If the tank is 50 ft long by 14 ft wide, what is the surface loading rate in gallons per day per square foot?

*Solution:*

$$\text{Surface loading rate} = \frac{\text{Flow (gal/day)}}{\text{Area (ft}^2\text{)}} = \frac{55{,}000\text{ gpd}}{50\text{ ft} \times 14\text{ ft}} = 179\text{ gpd/ft}^2$$

*Problem 4.10:* A sedimentation tanks is 88 ft long by 42 ft wide and receives a flow of 5.05 MGD. Calculate the surface loading rate in gallons per day per square foot.

*Solution:*

$$\text{Surface loading rate} = \frac{\text{Flow (gal/day)}}{\text{Area (ft}^2\text{)}} = \frac{5{,}050{,}000\text{ gpd}}{88\text{ ft} \times 42\text{ ft}} = 1366.3\text{ gpd/ft}^2$$

*Problem 4.11:* A circular clarifier has a diameter of 70 ft. If the flow to the clarifier is 3.5 MGD, what is the surface loading rate in gallons per day per square foot?

*Solution:*

$$\text{Surface loading rate} = \frac{\text{Flow (gal/day)}}{\text{Area (ft}^2)} = \frac{3{,}500{,}000 \text{ gpd}}{0.785 \times 70 \text{ ft} \times 70 \text{ ft}} = 909.92 \text{ gpd/ft}^2$$

*Problem 4.12:* A clarifier has a flow rate of 4800 gpm and a diameter of 80 ft. What is the surface loading rate in gallons per day per square foot?

*Solution:*

$$4800 \text{ gal/min} \times 1440 \text{ min/day} = 6{,}912{,}000 \text{ gpd}$$

$$\text{Surface loading rate} = \frac{\text{Flow (gal/day)}}{\text{Area (ft}^2)} = \frac{6{,}912{,}000 \text{ gpd}}{0.785 \times 80 \text{ ft} \times 80 \text{ ft}} = 1375.8 \text{ gpd/ft}^2$$

*Problem 4.13:* A clarifier with a diameter of 52 ft receives a flow of 2.085 MGD. What is the surface loading rate in gallons per day per square foot?

*Solution:*

$$\text{Surface loading rate} = \frac{\text{Flow (gal/day)}}{\text{Area (ft}^2)} = \frac{2{,}085{,}000 \text{ gpd}}{0.785 \times 52 \text{ ft} \times 52 \text{ ft}} = 982.3 \text{ gpd/ft}^2$$

*Problem 4.14:* What is the gpd/ft$^2$ overflow to a circular clarifier that has the following features:

Diameter = 80 ft
Flow = 1960 gpm

*Solution:*

$$1900 \text{ gal/min} \times 1440 \text{ min/day} = 2{,}822{,}400 \text{ gpd}$$

$$\text{Overflow} = \frac{\text{Flow (gal/day)}}{\text{Area (ft}^2)} = \frac{2{,}822{,}400 \text{ gpd}}{0.785 \times 80 \text{ ft} \times 80 \text{ ft}} = 562 \text{ gpd/ft}^2$$

*Problem 4.15:* A rectangular clarifier receives a flow of 5.5 MGD. The length of the clarifier is 96 ft, 6 in., and the width is 76 ft, 6 in. What is the surface loading rate in gallons per day per square foot?

*Solution:*

$$\text{Surface loading rate} = \frac{\text{Flow (gal/day)}}{\text{Area (ft}^2)} = \frac{5{,}500{,}00 \text{ gpd}}{96.5 \text{ ft} \times 96.5 \text{ ft}} = 744.95 \text{ gpd/ft}^2$$

*Problem 4.16:* A tank has a length of 110 ft, a width of 26 ft, and a depth of 12 ft. What is the weir length around the basin in feet?

*Solution:*

$$\text{Weir length} = (2 \times \text{Length}) + (2 \times \text{Width}) = (2 \times 110\text{ ft}) + (2 \times 26\text{ ft}) = 272\text{ ft}$$

*Problem 4.17:* A clarifier has a diameter of 80 ft and a depth of 14 ft. What is the length of the weir around the clarifier in feet?

*Solution:*

$$\text{Weir length} = 3.14 \times \text{Diameter (ft)} = 3.14 \times 80\text{ ft} = 251.2\text{ ft}$$

*Problem 4.18:* A sedimentation tank has a total of 140 ft of weir over which the water flows. What is the weir loading rate in gallons per day per foot of weir when the flow is 1.8 MGD?

*Solution:*

$$\text{Weir loading rate} = \frac{\text{Flow (gpd)}}{\text{Weir length (ft)}} = \frac{1{,}800{,}000\text{ gpd}}{140\text{ ft}} = 12{,}857.14\text{ gpd/ft}$$

*Problem 4.19:* The diameter of the weir in a circular clarifier is 90 ft. What is the weir loading rate (in gpd/ft) if the flow over the weir is 2.25 MGD?

*Solution:*

$$\text{Weir loading rate} = \frac{\text{Flow (gpd)}}{\text{Weir length (ft)}} = \frac{2{,}250{,}000\text{ gpd}}{90\text{ ft}} = 25{,}000\text{ gpd/ft}$$

*Problem 4.20:* A sedimentation tank has a total of 210 ft of weir over which the water flows. What is the weir overflow rate (in gpd/ft) when the flow is 2.4 MGD?

*Solution:*

$$\text{Weir loading rate} = \frac{\text{Flow (gpd)}}{\text{Weir length (ft)}} = \frac{2{,}400{,}000\text{ gpd}}{210\text{ ft}} = 11{,}429\text{ gpd/ft}$$

*Problem 4.21:* The diameter of the weir in a circular clarifier is 120 ft. The flow is 6.12 MGD. What is the weir loading rate in gallons per day per foot?

*Solution:*

$$\text{Weir loading rate} = \frac{\text{Flow (gpd)}}{3.14 \times \text{Weir diameter (ft)}} = \frac{6{,}120{,}000\text{ gpd}}{3.14 \times 210\text{ ft}} = 16{,}242.04\text{ gpd/ft}$$

*Problem 4.22:* A tank has a diameter of 48.8 ft. What is the gallons per day per foot of weir overflow when the tank receives 1,955,000 gpd?

*Solution:*

$$\text{Weir overflow rate} = \frac{\text{Flow (gpd)}}{3.14 \times \text{Weir diameter (ft)}} = \frac{1{,}955{,}000 \text{ gpd}}{3.14 \times 48.8 \text{ ft}} = 12{,}758.6 \text{ gpd/ft}$$

*Problem 4.23:* The flow rate to a particular clarifier is 530 gpm; the tank has a length of 32 ft and a width of 18 ft. What is the gpd/ft rate of weir overflow?

*Solution:*

$$530 \text{ gpm/min} \times 1440 \text{ min/day} = 763{,}200 \text{ gpd}$$

$$\text{Weir overflow rate} = \frac{763{,}200 \text{ gpd}}{(2 \times 32 \text{ ft}) + (2 \times 18 \text{ ft})} = \frac{763{,}200 \text{ gpd}}{64 + 36 \text{ ft}} = 7632 \text{ gpd/ft}$$

*Problem 4.24:* The weir in a basin measures 32 ft by 14 ft. What is the weir loading rate (in gpd/ft) when the flow is 1,096,000 gpd?

*Solution:*

$$\text{Weir loading rate} = \frac{1{,}096{,}000 \text{ gpd}}{(2 \times 32 \text{ ft}) + (2 \times 14 \text{ ft})} = \frac{1{,}096{,}000 \text{ gpd}}{64 + 28 \text{ ft}} = 11{,}913.04 \text{ gpd/ft}$$

*Problem 4.25:* What is the weir loading rate of a clarifier that is 52 ft, 4 in., by 46, ft, 3 in., and has an influent flow of 1.88 MGD?

*Solution:*

$$4/12 = 0.33 \text{ ft}; \quad 3/12 = 0.25 \text{ ft}$$

$$\text{Weir loading rate} = \frac{1{,}880{,}000 \text{ gpd}}{(2 \times 52.33 \text{ ft}) + (2 \times 46.25 \text{ ft})} = \frac{1{,}880{,}000}{104.66 + 92.50 \text{ ft}} = 9535 \text{ gpd/ft}$$

*Problem 4.26:* A tank has a length of 110 ft, a width of 26 ft, and a depth of 16 ft. What is the volume in gallons?

*Solution:*

$$\begin{aligned}\text{Volume} &= \text{Length (ft)} \times \text{Width (ft)} \times \text{Depth (ft)} \times 7.48 \text{ gal/ft}^3 \\ &= 110 \text{ ft} \times 26 \text{ ft} \times 16 \text{ ft} \times 7.48 \text{ gal/ft}^3 = 342{,}284.8 \text{ gal}\end{aligned}$$

*Problem 4.27:* A clarifier has a diameter of 80 ft and a depth of 10 ft. What is the volume of the clarifier in gallons?

*Solution:*

$$\text{Volume} = 0.785 \times D^2 \times \text{Depth (ft)} \times 7.48 \text{ gal/ft}^3$$
$$= 0.785 \times (80 \text{ ft})^2 \times 10 \text{ ft} \times 7.48 \text{ gal/ft}^3 = 375{,}795.2 \text{ gal}$$

*Problem 4.28:* A circular clarifier handles a flow of 0.8 MGD. The clarifier is 60 ft in diameter and 9 ft deep. What is the detention time in hours?

*Solution:*

$$\text{Detention time} = \frac{\text{Volume (gal)} \times 24 \text{ hr/day}}{\text{Flow (gpd)}}$$

$$= \frac{0.785 \times (60 \text{ ft})^2 \times 9 \text{ ft} \times 7.48 \text{ gal/ft}^3 \times 24 \text{ hr/day}}{800{,}000 \text{ gpd}} = 5.71 \text{ hr}$$

*Problem 4.29:* A clarifier is 72 ft long by 26 ft wide and is 12 ft deep. If the daily flow is 2,760,000 gpd, what is the detention time in the basin in hours?

*Solution:*

$$\text{Detention time} = \frac{\text{Volume (gal)} \times 24 \text{ hr/day}}{\text{Flow (gpd)}}$$

$$= \frac{72 \text{ ft} \times 26 \text{ ft} \times 12 \text{ ft} \times 7.48 \text{ gal/ft}^3 \times 24 \text{ hr/day}}{2{,}780{,}000 \text{ gpd}} = 1.45 \text{ hr}$$

*Problem 4.30:* What is the detention time in hours of a circular clarifier that receives a flow of 3200 gpm if the clarifier is 60 ft in diameter by 10 ft deep?

*Solution:*

$$3200 \text{ gal/min} \times 1440 \text{ min/day} = 4{,}608{,}000 \text{ gpd}$$

$$\text{Detention time} = \frac{\text{Volume (gal)} \times 24 \text{ hr/day}}{\text{Flow (gpd)}}$$

$$= \frac{0.785 \times (60 \text{ ft})^2 \times 10 \text{ ft} \times 7.48 \text{ gal/ft}^3 \times 24 \text{ hr/day}}{800{,}000 \text{ gpd}} = 5.71 \text{ hr}$$

*Problem 4.31:* A sedimentation tank is 62 ft long by 10 ft wide and has water to a depth of 10 ft. If the flow to the tank is 21,500 gph, what is the detention time in hours?

*Solution:*

$$21{,}500 \text{ gal/hr} \times 24 \text{ hr/day} = 516{,}000 \text{ gpd}$$

$$\text{Detention time} = \frac{\text{Volume (gal)} \times 24 \text{ hr/day}}{\text{Flow (gpd)}}$$

$$= \frac{62 \text{ ft} \times 10 \text{ ft} \times 10 \text{ ft} \times 7.48 \text{ gal/ft}^3 \times 24 \text{ hr/day}}{516{,}000 \text{ gpd}} = 2.16 \text{ hr}$$

*Problem 4.32:* A circular clarifier receives a flow of 900 gpm. If it has a diameter of 50 ft and a water depth of 8 ft, what is the detention time in hours?

*Solution:*

$$900 \text{ gal/min} \times 1440 \text{ min/day} = 1{,}296{,}000 \text{ gpd}$$

$$\text{Detention time} = \frac{\text{Volume (gal)} \times 24 \text{ hr/day}}{\text{Flow (gpd)}}$$

$$= \frac{0.785 \times (50 \text{ ft})^2 \times 8 \text{ ft} \times 7.48 \text{ gal/ft}^3 \times 24 \text{ hr/day}}{1{,}296{,}000 \text{ gpd}} = 2.17 \text{ hr}$$

*Problem 4.33:* A clear well is 60 ft long by 15 ft wide and has water to a depth of 6 ft. If the daily flow is 690 gpm, what is the detention time in minutes?

*Solution:*

$$690 \text{ gal/min} \times 1440 \text{ min/day} = 993{,}600 \text{ gpd}$$

$$\text{Detention time} = \frac{\text{Volume (gal)} \times 24 \text{ hr/day}}{\text{Flow (gpd)}}$$

$$= \frac{60 \text{ ft} \times 15 \text{ ft} \times 6 \text{ ft} \times 7.48 \text{ gal/ft}^3 \times 24 \text{ hr/day}}{993{,}600 \text{ gpd}} = 0.98 \text{ hr}$$

$$0.98 \text{ hr} \times 60 \text{ min/hr} = 58.8 \text{ min}$$

*Problem 4.34:* The flow to a sedimentation tank is 4.10 MGD. If the tank is 70 ft long by 30 ft wide, what is the surface loading rate in gallons per day per square foot?

*Solution:*

$$\text{Surface loading rate} = \frac{\text{Flow (gpd)}}{\text{Area (ft}^2\text{)}} = \frac{4{,}100{,}000 \text{ gpd}}{70 \text{ ft} \times 30 \text{ ft}} = 1952.38 \text{ gpd/ft}^2$$

*Problem 4.35:* The diameter of the weir in a circular clarifier is 120 ft. The flow is 5.50 MGD. What is the weir loading rate in gallons per day per square foot?

*Solution:*

$$\text{Weir loading rate} = \frac{\text{Flow (gal/day)}}{\text{Area (ft}^2\text{)}} = \frac{5{,}500{,}000 \text{ gpd}}{3.14 \times 120 \text{ ft}} = 14{,}596.6 \text{ gpd/ft}^2$$

*Problem 4.36:* A rectangular clarifier handles a flow of 3.42 MGD. The clarifier is 60 ft long by 30 ft wide and is 22 ft deep. What is the detention time in minutes?

*Solution:*

$$\text{Detention time} = \frac{\text{Volume (gal)} \times 24 \text{ hr/day} \times 60 \text{ min/day}}{\text{Flow (gpd)}}$$

$$= \frac{60 \text{ ft} \times 30 \text{ ft} \times 22 \text{ ft} \times 7.48 \text{ gal/ft}^3 \times 24 \text{ hr/day} \times 60 \text{ min/day}}{3{,}420{,}000 \text{ gpd}}$$

$$= 124.72 \text{ min}$$

*Problem 4.37:* A circular clarifier receives a flow of 3466.4 gpm. What is the detention time in the clarifier in hours? The clarifier has a diameter 63 ft and a depth of 20 ft.

*Solution:*

$$\text{Detention time} = \frac{\text{Volume (gal)} \times 24 \text{ hr/day}}{\text{Flow (gpd)}}$$

$$= \frac{0.785 \times (63 \text{ ft})^2 \times 20 \text{ ft} \times 7.48 \text{ gal/ft}^3 \times 24 \text{ hr/day}}{3466.4 \text{ gal/min} \times 1440 \text{ min/day}}$$

$$= 2.24 \text{ hr}$$

# 5 Filtration Calculations

I was always aware, I think, of the water in the soil, the way it travels from particle to particle, molecules adhering, clustering, evaporating, heating, cooling, freezing, rising upward to the surface and fogging the cool air or sinking downward, dissolving the nutrient and that, quick in everything it does, endlessly working and flowing.

**—J.A. Smiley (*A Thousand Acres*)**

## WATER FILTRATION

Water filtration is a physical process of separating suspended and colloidal particles from waste by passing the water through a granular material. The process of filtration involves straining, settling, and adsorption. As floc passes into the filter, the spaces between the filter grains become clogged, reducing this opening and increasing removal. Some material is removed merely because it settles on a media grain. One of the most important processes is adsorption of the floc onto the surface of individual filter grains. In addition to removing silt and sediment, floc, algae, insect larvae, and any other large elements, filtration also contributes to the removal of bacteria and protozoans such as *Giardia lamblia* and *Cryptosporidium*. Some filtration processes are also used for iron and manganese removal.

The Surface Water Treatment Rule (SWTR) specifies four filtration technologies, although the SWTR also allows the use of alternative filtration technologies, such as cartridge filters. The four filtration technologies are slow sand (see Figure 5.1) or rapid sand filtration, pressure filtration, diatomaceous earth filtration, and direct filtration. Of these, all but rapid sand filtration are commonly employed in small water systems that use filtration. Each type of filtration system has advantages and disadvantages. Regardless of the type of filter, however, filtration involves the processes of *straining* (where particles are captured in the small spaces between filter media grains), *sedimentation* (where the particles land on top of the grains and stay there), and *adsorption* (where a chemical attraction occurs between the particles and the surface of the media grains.

## FLOW RATE THROUGH A FILTER

Flow rate in gallons per minute (gpm) through a filter can be determined by simply converting the gallon per day (gpd) flow rate, as indicated on the flow meter. The flow rate in gallons per minute can be calculated by taking the meter flow rate (gpd) and dividing by 1440 min/day, as shown in Equation 5.1.

$$\text{Flow rate (gpm)} = \frac{\text{Flow rate (gpd)}}{1440 \text{ min/day}} \tag{5.1}$$

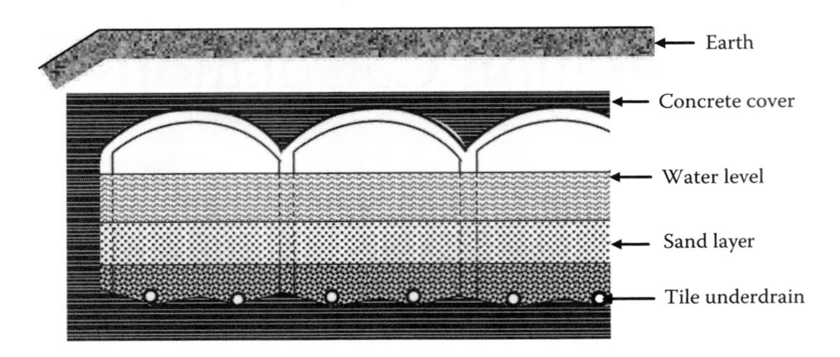


**FIGURE 5.1** Slow sand filter.

### ■ Example 5.1

*Problem:* The flow rate through a filter is 4.25 million gallons per day (MGD). What is this flow rate expressed as gpm?

*Solution:*

$$\text{Flow rate} = \frac{\text{Flow rate (gpd)}}{\text{1440 min/day}} = \frac{\text{4.25 MGD}}{\text{1440 min/day}} = \frac{\text{4,250,000 gpd}}{\text{1440 min/day}} = \text{2951 gpm}$$

### ■ Example 5.2

*Problem:* During a 70-hour filter run, 22.4 million gallons of water are filtered. What is the average flow rate through the filter (in gpm) during this filter run?

*Solution:*

$$\text{Flow rate} = \frac{\text{Total gallons produced}}{\text{Filter run (min)}} = \frac{\text{22,400,000 gal}}{\text{70 hr} \times \text{60 min/hr}} = \text{5333 gpm}$$

### ■ Example 5.3

*Problem:* At an average flow rate of 4000 gpm, how long of a filter run (in hours) would be required to produce 25 MG of filtered water?

*Solution:*

$$\text{Flow rate} = \frac{\text{Total gallons produced}}{\text{Filter run (min)}}$$

$$4000 = \frac{\text{25,000,000 gal}}{x \text{ hr} \times \text{60 min/hr}}$$

$$x = \frac{\text{25,000,000 gal}}{4000 \times 60} = \text{104 hr}$$

## ■ Example 5.4

*Problem:* A filter box is 20 ft by 30 ft (including the sand area). If the influent valve is shut, the water drops 3 in. per minute. What is the rate of filtration in MGD?

*Solution:* We know that the filter box is 20 ft by 30 ft and that the water drops 3 in. per minute.

> ***Note:*** The best way to perform calculations for this type of problem is step by step, breaking down the problem into what is given and what is to be found.

First find the volume of water passing through the filter:

$$\text{Volume} = \text{Area} \times \text{Height}$$

$$\text{Area} = \text{Width} \times \text{Length} = 20 \text{ ft} \times 30 \text{ ft} = 600 \text{ ft}^2$$

Divide 3 in. by 12 to find feet: $3 \div 12 = 0.25$ ft.

$$\text{Volume} = 600 \text{ ft}^2 \times 0.25 \text{ ft} = 150 \text{ ft}^3 \text{ of water passing through filter in 1 min}$$

Now convert cubic feet to gallons:

$$150 \text{ ft}^3 \times 7.48 \text{ gal/ft}^3 = 1122 \text{ gal/min}$$

The problem asks for the rate of filtration in MGD. To find MGD, multiply the number of gallons per minute by the number of minutes per day:

$$1122 \text{ gal/min} \times 1440 \text{ min/day} = 1.62 \text{ MGD}$$

## ■ Example 5.5

*Problem:* The influent valve to a filter is closed for 5 minutes. During this time, the water level in the filter drops 0.8 ft (10 in.). If the filter is 45 ft long by 15 ft wide, what is the gpm flow rate through the filter? The water drop is 0.16 ft/min.

*Solution:* First calculate the cubic feet per minute (cfm) flow rate using the $Q = A \times V$ equation:

$$Q = \text{Length (ft)} \times \text{Width (ft)} \times \text{Velocity (ft/min)}$$

$$= 45 \text{ ft} \times 15 \text{ ft} \times 0.16 \text{ ft/min}$$

$$= 108 \text{ cfm}$$

Then convert the cfm flow rate to the gpm flow rate:

$$\text{Flow} = 108 \text{ cfm} \times 7.48 \text{ gal/ft}^3 = 808 \text{ gpm}$$

## FILTRATION RATE

One measure of filter production is the filtration rate (which generally ranges from 2 to 10 gpm/ft$^2$). Along with filter run time, it provides valuable information for operation of filters. The rate is the number of gallons of water filtered per minute through each square foot of filter area. The filtration rate is determined using Equation 5.2:

$$\text{Filtration rate (gpm/ft}^2\text{)} = \frac{\text{Flow rate (gpm)}}{\text{Filter surface area (ft}^2\text{)}} \tag{5.2}$$

### ■ Example 5.6

*Problem:* A filter 18 ft by 22 ft receives a flow of 1750 gpm. What is the filtration rate in gpm/ft$^2$?

*Solution:*

$$\text{Filtration rate} = \frac{\text{Flow rate (gpm)}}{\text{Filter surface area (ft}^2\text{)}} = \frac{1750 \text{ gpm}}{18 \text{ ft} \times 22 \text{ ft}} = 4.4 \text{ gpm/ft}^2$$

### ■ Example 5.7

*Problem:* A filter 28 ft long by 18 ft wide treats a flow of 3.5 MGD. What is the filtration rate in gpm/ft$^2$?

*Solution:*

$$\text{Flow rate} = \frac{\text{Flow rate (gpd)}}{1440 \text{ min/day}} = \frac{3{,}500{,}000 \text{ (gpd)}}{1440 \text{ min/day}} = 2431 \text{ gpm}$$

$$\text{Filtration rate} = \frac{\text{Flow rate (gpm)}}{\text{Filter surface area (ft}^2\text{)}} = \frac{2431 \text{ gpm}}{28 \text{ ft} \times 18 \text{ ft}} = 4.8 \text{ gpm/ft}^2$$

### ■ Example 5.8

*Problem:* A filter 45 ft long by 20 ft wide produces a total of 18 MG during a 76-hour filter run. What is the average filtration rate (in gpm/ft$^2$) for this filter run?

*Solution:*

$$\text{Flow rate} = \frac{\text{Total gallons produced}}{\text{Filter run (min)}} = \frac{18{,}000{,}000 \text{ gal}}{76 \text{ hr} \times 60 \text{ min/hr}} = 3947 \text{ gpm}$$

$$\text{Filtration rate} = \frac{\text{Flow rate (gpm)}}{\text{Filter surface area (ft}^2\text{)}} = \frac{3947 \text{ gpm}}{45 \text{ ft} \times 20 \text{ ft}} = 4.4 \text{ gpm/ft}^2$$

■ **EXAMPLE 5.9**

*Problem:* A filter is 40 ft long by 20 ft wide. During a test of flow rate, the influent valve to the filter is closed for 6 minutes. The water level drop during this period is 16 in. What is the filtration rate for the filter in gpm/ft$^2$?

*Solution:* First calculate gpm flow rate, using the $Q = A \times V$ equation:

$$Q = \text{Length (ft)} \times \text{Width (ft)} \times \text{Velocity (ft/min)} \times 7.48 \text{ gal/ft}^3$$

$$= 40 \text{ ft} \times 20 \text{ ft} \times (1.33 \text{ ft/6 min}) \times 7.48 \text{ gal/ft}^3 = 1316 \text{ gpm}$$

Then calculate filtration rate:

$$\text{Filtration rate} = \frac{\text{Flow rate (gpm)}}{\text{Filter surface area (ft}^2)} = \frac{1316 \text{ gpm}}{40 \text{ ft} \times 20 \text{ ft}} = 1.657 \text{ (1.7) gpm/ft}^2$$

## UNIT FILTER RUN VOLUME

The unit filter run volume (UFRV) calculation measures the total gallons passing through each square foot of filter surface area during an entire filter run. This calculation is used to compare and evaluate filter runs. UFRVs are usually at least 5000 gal/ft$^2$ and generally in the range of 10,000 gpd/ft$^2$. The UFRV value will begin to decline as the performance of the filter begins to deteriorate. The equation to be used in these calculations is shown below:

$$\text{UFRV} = \frac{\text{Total gallons filtered}}{\text{Filter surface area (ft}^2)} \tag{5.3}$$

■ **EXAMPLE 5.10**

*Problem:* The total water filtered during a filter run (between backwashes) is 2,220,000 gal. If the filter is 18 ft by 18 ft, what is the unit filter run volume in gal/ft$^2$?

*Solution:*

$$\text{UFRV} = \frac{\text{Total gallons filtered}}{\text{Filter surface area (ft}^2)} = \frac{2{,}220{,}000 \text{ gal}}{18 \text{ ft} \times 18 \text{ ft}} = 6852 \text{ gal/ft}^2$$

■ **EXAMPLE 5.11**

*Problem:* The total water filtered during a filter run is 4,850,000 gal. If the filter is 28 ft by 18 ft, what is the unit filter run volume in gal/ft$^2$?

*Solution:*

$$\text{UFRV} = \frac{\text{Total gallons filtered}}{\text{Filter surface area (ft}^2)} = \frac{4{,}850{,}000 \text{ gal}}{28 \text{ ft} \times 18 \text{ ft}} = 9623 \text{ gal/ft}^2$$

Equation 5.3 can be modified as shown in Equation 5.4 to calculate the unit filter run volume given the filtration rate and filter run data.

$$\text{UFRV} = \text{Filtration rate (gpm/ft}^2) \times \text{Filter run time (min)} \tag{5.4}$$

### ■ Example 5.12

*Problem:* The average filtration rate for a filter was determined to be 2.0 gpm/ft². If the filter run time was 4250 minutes, what was the unit filter run volume in gal/ft²?

*Solution:*

$$\text{UFRV} = \text{Filtration rate (gpm/ft}^2) \times \text{Filter run time (min)}$$

$$= 2\ \text{gpm/ft}^2 \times 4250\ \text{min} = 8500\ \text{gal/ft}^2$$

The problem indicates that at an average filtration rate of 2.0 gal entering each square foot of filter each minute, the total gallons entering during the total filter run is 4250 times that amount.

### ■ Example 5.13

*Problem:* The average filtration rate during a particular filter run was determined to be 3.2 gpm/ft². If the filter run time was 61.0 hr, what was the UFRV (in gal/ft²) for the filter run?

*Solution:*

$$\text{UFRV} = \text{Filtration rate (gpm/ft}^2) \times \text{Filter run time (min)}$$

$$= 3.2\ \text{gpm/ft}^2 \times 61\ \text{hr} \times 60\ \text{min/hr} = 11{,}712\ \text{gal/ft}^2$$

## BACKWASH RATE

In filter backwashing, one of the most important operational parameters to be determined is the amount of water (in gal) required for each backwash. This amount depends on the design of the filter and the quality of the water being filtered. The actual washing typically lasts 5 to 10 minutes and uses 1 to 5% of the flow produced.

### ■ Example 5.14

*Problem:* A filter is 30 ft long by 20 ft wide and the depth of the filter media is 24 in. Assuming that a backwash rate of 15 gal/ft²/min is recommended, and 10 minutes of backwash are required, calculate the amount of water (in gal) required for each backwash.

*Solution:* Determine the amount of water required:

Area of filter = 30 ft × 20 ft = 600 ft²
Gallons of water used = 15 gal/ft²/min × 10 min = 150 gal/ft²
Gallons required = 150 gal/ft² × 600 ft² = 90,000 gal

Typically, backwash rates will range from 10 to 25 gpm/ft$^2$. The backwash rate is determined by using Equation 5.5:

$$\text{Backwash rate (gpm/ft}^2) = \frac{\text{Flow rate (gpm)}}{\text{Filter surface area (ft}^2)} \tag{5.5}$$

## ■ Example 5.15

*Problem:* A filter 30 ft by 10 ft has a backwash rate of 3120 gpm. What is the backwash rate, in gpm/ft$^2$?

*Solution:*

$$\text{Backwash rate} = \frac{\text{Flow rate (gpm)}}{\text{Filter surface area (ft}^2)} = \frac{3120\text{ gpm}}{30\text{ ft} \times 10\text{ ft}} = 10.4\text{ gpm/ft}^2$$

## ■ Example 5.16

*Problem:* A filter 20 ft long by 20 ft wide has a backwash flow rate of 4.85 MGD. What is the filter backwash rate, in gpm/ft$^2$?

*Solution:*

$$\frac{4{,}850{,}000\text{ gpd}}{1440\text{ min/day}} = 3368\text{ gpm}$$

$$\text{Backwash rate} = \frac{\text{Flow rate (gpm)}}{\text{Filter surface area (ft}^2)} = \frac{3368\text{ gpm}}{20\text{ ft} \times 20\text{ ft}} = 8.42\text{ gpm/ft}^2$$

## Backwash Rise Rate

Backwash rate is occasionally measured as the upward velocity of the water during backwashing, expressed as in./min rise. To convert from a gpm/ft$^2$ backwash rate to an in./min rise rate, use either Equation 5.6 or Equation 5.7:

$$\text{Backwash rate (in./min)} = \frac{\text{Backwash rate (gpm/ft}^2) \times 12\text{ in./ft}}{7.48\text{ gal/ft}^3} \tag{5.6}$$

$$\text{Backwash rate (in./min)} = \text{Backwash rate (gpm/ft}^2) \times 1.6 \tag{5.7}$$

## ■ Example 5.17

*Problem:* A filter has a backwash rate of 16 gpm/ft$^2$. What is this backwash rate expressed as an in./min rise rate?

*Solution:*

$$\text{Backwash rate} = \frac{\text{Backwash rate (gpm/ft}^2) \times 12 \text{ in./ft}}{7.48 \text{ gal/ft}^3}$$

$$= \frac{16 \text{ gpm/ft}^2 \times 12 \text{ in./ft}}{7.48 \text{ gal/ft}^3} = 25.7 \text{ in./min}$$

### ■ Example 5.18

*Problem:* A filter 22 ft long by 12 ft wide has a backwash rate of 3260 gpm. What is this backwash rate expressed as an in./min rise?

*Solution:* First calculate the backwash rate as gpm/ft²:

$$\text{Backwash rate} = \frac{\text{Flow rate (gpm)}}{\text{Filter area (ft}^2)} = \frac{3260 \text{ gpm}}{22 \text{ ft} \times 12 \text{ ft}} = 12.3 \text{ gpm/ft}^2$$

Then convert the gpm/ft² to an in./min rise rate:

$$\text{Backwash rate} = \frac{12.3 \text{ gpm/ft}^2 \times 12 \text{ in./ft}}{7.48 \text{ gal/ft}^3} = 19.7 \text{ in./min}$$

## VOLUME OF BACKWASH WATER REQUIRED

To determine the volume of water required for backwashing, we must know both the desired backwash flow rate (gpm) and the duration of backwash (min):

$$\text{Backwash volume (gal)} = \text{Backwash rate (gpm)} \times \text{Duration of backwash (min)} \quad (5.8)$$

### ■ Example 5.19

*Problem:* For a backwash flow rate of 9000 gpm and a total backwash time of 8 minutes, how many gallons of water will be required for backwashing?

*Solution:*

$$\text{Backwash volume} = \text{Backwash rate (gpm)} \times \text{Duration of backwash (min)}$$

$$= 9000 \text{ gpm} \times 8 \text{ min} = 72{,}000 \text{ gal}$$

### ■ Example 5.20

*Problem:* How many gallons of water would be required to provide a backwash flow rate of 4850 gpm for a total of 5 minutes?

*Solution:*

$$\text{Backwash volume} = \text{Backwash rate (gpm)} \times \text{Duration of backwash (min)}$$

$$= 4850 \text{ gpm} \times 7 \text{ min} = 33{,}950 \text{ gal}$$

## REQUIRED DEPTH OF BACKWASH WATER TANK

The required depth of water in the backwash water tank is determined from the volume of water required for backwashing. To make this calculation, use Equation 5.9:

$$\text{Backwash volume (gal)} = 0.785 \times D^2 \times \text{Depth (ft)} \times 7.48\ \text{gal/ft}^3 \tag{5.9}$$

### ■ EXAMPLE 5.21

*Problem:* The volume of water required for backwashing has been calculated to be 85,000 gal. What is the required depth of water in the backwash water tank to provide this amount of water if the diameter of the tank is 60 ft?

*Solution:* Use the volume equation for a cylindrical tank, fill in the known data, and then solve for $x$:

$$\text{Backwash volume} = 0.785 \times D^2 \times \text{Depth (ft)} \times 7.48\ \text{gal/ft}^3$$

$$85{,}000\ \text{gal} = 0.785 \times (60\ \text{ft})^2 \times x\ \text{ft} \times 7.48\ \text{gal/ft}^3$$

$$x = \frac{85{,}000\ \text{gal}}{0.785 \times 3600\ \text{ft}^2 \times 7.48\ \text{gal/ft}^3}$$

$$= 4\ \text{ft}$$

### ■ EXAMPLE 5.22

*Problem:* A total of 66,000 gal of water will be required for backwashing a filter at a rate of 8000 gpm for a 9-min period. What depth of water is required if the backwash tank has a diameter of 50 ft?

*Solution:* Use the volume equation for cylindrical tanks:

$$\text{Backwash volume} = 0.785 \times D^2 \times \text{Depth (ft)} \times 7.48\ \text{gal/ft}^3$$

$$66{,}000\ \text{gal} = 0.785 \times (50\ \text{ft})^2 \times x\ \text{ft} \times 7.48\ \text{gal/ft}^3$$

$$x = \frac{66{,}000\ \text{gal}}{0.785 \times 2500\ \text{ft}^2 \times 7.48\ \text{gal/ft}^3} = 4.5\ \text{ft}$$

## BACKWASH PUMPING RATE

The desired backwash pumping rate (gpm) for a filter depends on the desired backwash rate in gpm/ft² and the ft² area of the filter. The backwash pumping rate in gpm can be determined by using Equation 5.10.

$$\text{Backwash pumping rate} = \text{Desired backwash rate (gpm/ft}^2) \times \text{Filter area (ft}^2) \tag{5.10}$$

■ **Example 5.23**

*Problem:* A filter is 25 ft long by 20 ft wide. If the desired backwash rate is 22 gpm/ft², what backwash pumping rate (in gpm) will be required?

*Solution:* The desired backwash flow through each square foot of filter area is 20 gpm. The total gpm flow through the filter is therefore 20 gpm times the entire square foot area of the filter:

$$\begin{aligned}\text{Backwash pumping rate} &= \text{Desired backwash rate (gpm/ft}^2) \times \text{Filter area (ft}^2)\\ &= 20\text{ gpm/ft}^2 \times 25\text{ ft} \times 20\text{ ft} = 10{,}000\text{ gpm}\end{aligned}$$

■ **Example 5.24**

*Problem:* The desired backwash-pumping rate for a filter is 12 gpm/ft². If the filter is 20 ft long by 20 ft wide, what backwash pumping rate (in gpm) will be required?

*Solution:*

$$\begin{aligned}\text{Backwash pumping rate} &= \text{Desired backwash rate (gpm/ft}^2) \times \text{Filter area (ft}^2)\\ &= 12\text{ gpm/ft}^2 \times 20\text{ ft} \times 20\text{ ft} = 4800\text{ gpm}\end{aligned}$$

## PERCENT PRODUCT WATER USED FOR BACKWASHING

Along with measuring filtration rate and filter run time, another aspect of filter operation that is monitored for filter performance is the percent of product water used for backwashing. The equation for percent of product water used for backwashing calculations used is shown below:

$$\text{Backwash water (\%)} = \frac{\text{Backwash water (gal)}}{\text{Water filtered (gal)}} \times 100 \tag{5.11}$$

■ **Example 5.25**

*Problem:* A total of 18,100,000 gal of water were filtered during a filter run. If 74,000 gal of this product water were used for backwashing, what percent of the product water was used for backwashing?

*Solution:*

$$\text{Backwash water} = \frac{\text{Backwash water (gal)}}{\text{Water filtered (gal)}} \times 100 = \frac{74{,}000\text{ gal}}{18{,}100{,}000\text{ gal}} \times 100 = 0.4\%$$

■ **Example 5.26**

*Problem:* A total of 11,400,000 gal of water are filtered during a filter run. If 48,500 gal of product water are used for backwashing, what percent of the product water is used for backwashing?

*Solution:*

$$\text{Backwash water} = \frac{\text{Backwash water (gal)}}{\text{Water filtered (gal)}} \times 100 = \frac{48{,}500 \text{ gal}}{11{,}400{,}000 \text{ gal}} \times 100 = 0.43\%$$

## PERCENT MUD BALL VOLUME

Mud balls are heavier deposits of solids near the top surface of the medium that break into pieces during backwash, resulting in spherical accretions (usually less than 12 inches in diameter) of floc and sand. The presence of mud balls in the filter media is checked periodically. The principal objection to mud balls is that they diminish the effective filter area. To calculate the percent mud ball volume, we use Equation 5.12:

$$\%\ \text{Mud ball volume} = \frac{\text{Mud ball volume (mL)}}{\text{Total sample volume (mL)}} \times 100 \tag{5.12}$$

### ■ Example 5.27

*Problem:* A 3350-mL sample of filter media was taken for mud ball evaluation. The volume of water in the graduated cylinder rose from 500 mL to 525 mL when mud balls were placed in the cylinder. What is the percent mud ball volume of the sample?

*Solution:* First, determine the volume of mud balls in the sample:

$$525 \text{ mL} - 500 \text{ mL} = 25 \text{ mL}$$

Then calculate the percent mud ball volume:

$$\%\ \text{Mud ball volume} = \frac{\text{Mud ball volume (mL)}}{\text{Total sample volume (mL)}} \times 100 = \frac{25 \text{ mL}}{3350 \text{ mL}} \times 100 = 0.70\%$$

### ■ Example 5.28

*Problem:* A filter is tested for the presence of mud balls. The mud ball sample has a total sample volume of 680 mL. Five samples were taken from the filter. When the mud balls were placed in 500 mL of water, the water level rose to 565 mL. What is the percent mud ball volume of the sample?

*Solution:* The mud ball volume is the volume the water rose:

$$565 \text{ mL} - 500 \text{ mL} = 65 \text{ mL}$$

Because five samples of media were taken, the total sample volume is 5 times the sample volume: $5 \times 680$ mL = 3400 mL.

$$\%\ \text{Mud ball volume} = \frac{\text{Mud ball volume (mL)}}{\text{Total sample volume (mL)}} \times 100 = \frac{65 \text{ mL}}{3400 \text{ mL}} \times 100 = 1.9\%$$

## FILTER BED EXPANSION

In addition to backwash rate, it is also important to expand the filter media during the wash to maximize the removal of particles held in the filter or by the media; that is, the efficiency of the filter wash operation depends on the expansion of the sand bed. Bed expansion is determined by measuring the distance from the top of the unexpanded media to a reference point (e.g., top of the filter wall) and from the top of the expanded media to the same reference. A proper backwash rate should expand the filter 20 to 25%. Percent bed expansion is given by dividing the bed expansion by the total depth of expandable media (i.e., media depth less support gravels) and multiplied by 100 as follows:

Expanded measurement = Depth to top of media during backwash (in.)
Unexpanded measurement = Depth to top of media before backwash (in.)
Bed expansion = Unexpanded measurement (in.) – Expanded measurement (in.)

$$\% \text{ Bed expansion} = \frac{\text{Bed expansion measurement (in.)}}{\text{Total depth of expandable media (in.)}} \times 100 \qquad (5.13)$$

### ■ EXAMPLE 5.29*

*Problem:* The backwashing practices for a filter with 30 in. of anthracite and sand are being evaluated. While at rest, the distance from the top of the media to the concrete floor surrounding the top of filter is measured to be 41 inches. After the backwash has been started and the maximum backwash rate is achieved, a probe containing a white disk is slowly lowered into the filter bed until anthracite is observed on the disk. The distance from the expanded media to the concrete floor is measured to be 34.5 inches. What is the percent bed expansion?

*Solution:*

Unexpanded measurement = 41 in.
Expanded measurement = 34.5 in.
Bed expansion = 41 in. – 34.5 in. = 6.5 in.
% Bed expansion = (6.5 in./30 in.) × 100 = 22%

## FILTER LOADING RATE

Filter loading rate is the flow rate of water applied to the unit area of the filter. It is the same value as the flow velocity approaching the filter surface and can be determined by using Equation 5.14.

$$u = Q/A \qquad (5.14)$$

* Adapted from USEPA, *EPA Guidance Manual Turbidity Provisions*, U.S. Environmental Protection Agency, Washington, DC, 1999, pp. 5-1–5-12.

where:

$u$ = Loading rate ($m^3/m^2 \cdot$ day, $L/m^2 \cdot$ min, or $gpm/ft^2$).
Q = Flow rate ($m^3$/day, $ft^3$/day, or gpm).
A = Surface area of filter ($m^2$ or $ft^2$).

Filters are classified as slow sand filters, rapid sand filters, and high-rate sand filters on the basis of loading rate. Typically, the loading rate for rapid sand filters is 120 $m^3/m^2 \cdot$ day (83 $L/m^2 \cdot$ min or 2 $gpm/ft^2$). The loading rate may be up to five times this rate for high-rate filters.

### ■ Example 5.30

*Problem:* A sanitation district is to install rapid sand filters downstream of the clarifiers. The design loading rate is selected to be 150 $m^3/m^2$. The design capacity of the waterworks is 0.30 $m^3$/sec (6.8 MGD). The maximum surface per filter is limited to 45 $m^2$. Design the number and size of filters and calculate the normal filtration rate.

*Solution:* Determine the total surface area required:

$$A = \frac{Q}{u} = \frac{0.30 \text{ m}^3\text{/sec} \times 86{,}400 \text{ sec/day}}{150 \text{ m}^3\text{/m}^2 \cdot \text{day}} = \frac{25{,}920 \text{ m}^3\text{/day}}{150 \text{ m}^3\text{/m}^2 \cdot \text{day}} = 173 \text{ m}^2$$

Determine the number of filters:

$$\frac{173 \text{ m}^2}{45 \text{ m}} = 3.8$$

Select four filters. The surface area ($A$) for each filter is

$$A = 173 \text{ m}^2 \div 4 = 43.25 \text{ m}^2$$

We can use 6 m × 7 m, 6.4 m × 7 m, or 6.42 m × 7 m. If a filter 6 m × 7 m is installed, the normal filtration rate is

$$u = \frac{Q}{A} = \frac{0.30 \text{ m}^3\text{/sec} \times 86{,}400 \text{ sec/day}}{4 \times 6 \text{ m} \times 7 \text{ m}} = 154.3 \text{ m}^2\text{/m}^3 \cdot \text{day}$$

## FILTER MEDIUM SIZE

Filter medium grain size has an important effect on the filtration efficiency and on backwashing requirements for the medium. The actual medium selected is typically determined by performing a grain size distribution analysis—sieve size and percentage passing by weight relationships are plotted on logarithmic–probability paper. The most common parameters used in the United States to characterize the filter medium are effective size (ES) and uniformity coefficient (UC) of medium size distribution. The ES is the grain size for which 10% of the grains are smaller by weight; it is often abbreviated by $d_{10}$. The UC is the ratio of the 60th percentile ($d_{60}$) to the 10th percentile. The 90th percentile ($d_{90}$) is the size for which 90% of the grains are smaller by weight. The $d_{90}$ size is used for computing the required filter backwash rate for a filter medium.

Values of $d_{10}$, $d_{60}$, and $d_{90}$ can be read from an actual sieve analysis curve. If such a curve is not available and if a linear logarithmic–probability plot is assumed, the values can be interrelated by Equation 5.15 (Cleasby, 1990):

$$d_{90} = d_{10} \times (10^{1.67 \log UC}) \tag{5.15}$$

### ■ Example 5.31

*Problem:* A sieve analysis curve of a typical filter sand gives $d_{10} = 0.52$ mm and $d_{60} = 0.70$ mm. What are the uniformity coefficient and $d_{90}$?

*Solution:*

$$UC = d_{60}/d_{10} = 0.70 \text{ mm}/0.52 \text{ mm} = 1.35$$

Find $d_{90}$ using Equation 5.15:

$$d_{90} = d_{10} \times (10^{1.67 \log UC}) = 0.52 \text{ mm} \times (10^{1.67 \log 1.35}) = 0.52 \text{ mm} \times (10^{0.218}) = 0.86 \text{ mm}$$

## MIXED MEDIA

Recently, an innovation in filtering systems has offered a significant improvement and economic advantage to rapid rate filtration: the mixed media filter bed. Mixed media filter beds offer specific advantages in specific circumstances but will give excellent operating results at a filtering rate of 5 gal/ft²/min. Moreover, the mixed media filtering unit is more tolerant of higher turbidities in the settled water. For improved process performance, activated carbon or anthracite is added on the top of the sand bed. The approximate specific gravities ($s$) of ilmenite (Chavara, <60% $TiO_2$), silica sand, anthracite, and water are 4.2, 2.6, 1.5, and 1.0, respectively. The economic advantage of the mixed bed media filter is based on filter area; it will safely produce 2-1/2 times as much filtered water as a rapid sand filter.

When settling velocities are equal, the particle sizes for media of different specific gravity can be computed by using Equation 5.16:

$$\frac{d_1}{d_2} = \left( \frac{s_2 - s}{s_1 - s} \right)^{2/3} \tag{5.16}$$

where

$d_1$, $d_2$ = Diameter of particles 1 and 2, respectively.

$s$, $s_1$, $s_2$ = Specific gravity of water and particles 1 and 2, respectively.

■ **Example 5.32**

*Problem:* Estimate the particle size of ilmenite sand (specific gravity = 4.2) that has the same settling velocity of silica sand 0.60 mm in known diameter (specific gravity = 2.6)

*Solution:* Find the diameter of ilmenite sand by using Equation 5.16:

$$\frac{d_1}{d_2} = \left(\frac{s_2 - s}{s_1 - s}\right)^{2/3} = 0.6 \text{ mm} \times \left(\frac{2.6-1}{4.2-1}\right)^{2/3} = 0.38 \text{ mm settling size}$$

## HEAD LOSS FOR FIXED BED FLOW

When water is pumped upward through a bed of fine particles at a very low flow rate the water percolates through the pores (void spaces) without disturbing the bed. This is a fixed bed process. The head loss (pressure drop) through a clean granular-media filter is generally less than 0.9 m (3 ft). With the accumulation of impurities, head loss gradually increases until the filter is backwashed. The Kozeny equation, shown below, is typically used for calculating head loss through a clean fixed bed flow filter:

$$\frac{h}{L} = \frac{k\mu(1-\varepsilon)^2}{gp\varepsilon^3} \times \left(\frac{A}{V}\right)^2 \times v \tag{5.17}$$

where

$h$ = Head loss in filter depth (m or ft).
$L$ = Filter depth (m or ft).
$k$ = Dimensionless Kozeny constant (5 for sieve openings, 6 for size of separation).
$\mu$ = Absolute viscosity of water (N sec/m$^2$ or lb sec/ft$^2$).
$\varepsilon$ = Porosity (dimensionless).
$g$ = Acceleration of gravity = 9.81 m/sec or 32.2 ft/sec.
$p$ = Density of water (kg/m$^3$ or lb/ft$^3$).
$A/V$ = Grain surface area per unit volume of grain.
= Specific surface $S$ (shape factor = 6.0 to 7.7).
= $6/d$ for spheres.
= $6/\psi d_{eq}$ for irregular grains.
$\psi$ = Grain sphericity or shape factor.
$d_{eq}$ = Grain diameter of spheres of equal volume
$v$ = Filtration (superficial) velocity (m/sec or fps).

■ **Example 5.33**

*Problem:* A dual-medium filter is composed of 0.3-m anthracite (mean size, 2.0 mm) that is placed over a 0.6-m layer of sand (mean size, 0.7 mm) with a filtration rate of 9.78 m/hr. Assume that the grain sphericity is $\psi$ = 0.75 and the porosity for both is 0.42. Although normally taken from the appropriate table at 15°C, we provide the head loss data for the filter at 1.131 × 10$^{-6}$ m$^2$/sec. (Adapted from Rose, 1951; Fair et al., 1968.)

*Solution:* Determine the head loss through the anthracite layer using the Kozeny equation (Equation 5.17):

$$\frac{h}{L} = \frac{k\mu(1-\varepsilon)^2}{gp\varepsilon^3} \times \left(\frac{A}{V}\right)^2 \times v$$

where
$L = 0.3$ m
$k = 6$
$g = 9.81$ m/s$^2$
$\mu = 1.131 \times 10^{-6}$ m$^2$/sec (from the appropriate table)
$\varepsilon = 0.4$
$A/V = 6/0.75d = 8/d = 8/0.002$
$v = 9.78$ m/hr $= 0.00272$ m/sec

Then,

$$h = 6 \times \frac{1.131 \times 10^{-6}}{9.81} \times \frac{(1-0.4)^2}{0.4^3} \times \left(\frac{8}{0.002}\right)^2 \times 0.00272 \times 0.3 = 0.0508 \text{ m}$$

Now compute the head loss passing through the sand. Use the data above, except insert:

$k = 5$
$d = 0.0007$ m
$L = 0.6$ m

$$h = 5 \times \frac{1.131 \times 10^{-6}}{9.81} \times \frac{0.6^2}{0.4^3} \times \left(\frac{8}{0.0007}\right)^2 \times 0.00272 \times 0.6 = 0.6918 \text{ m}$$

Now compute the total head loss:

$$h = 0.0508 \text{ m} + 0.6918 \text{ m} = 0.7426 \text{ m}$$

## HEAD LOSS THROUGH A FLUIDIZED BED

If the upward water flow rate through a filter bed is very large, the bed mobilizes pneumatically and may be swept out of the process vessel. At an intermediate flow rate the bed expands and is in what we call an *expanded* state. In the fixed bed, the particles are in direct contact with each other, supporting each other's weight. In the expanded bed, the particles have a mean free distance between particles and the drag force of the water supports the particles. The expanded bed has some of the properties of the water (i.e., of a fluid) and is called a fluidized bed (Chase, 2002). Simply, *fluidization* is defined as upward flow through a granular filter bed at sufficient velocity to suspend the grains in the water. Minimum fluidizing velocity ($U_{mf}$) is the superficial fluid velocity required to begin fluidization; it is important in determining the required minimum backwashing flow rate. The $U_{mf}$ equation including the near constants (over a wide range of particles) 33.7 and 0.0408, but excluding porosity of fluidization and shape factor, is as follows (Wen and Yu, 1966):

$$U_{mf} = \frac{\mu}{pd_{eq}} \times (1135.69 + 0.0408 \times G_n)^{0.5} - \frac{33.7\mu}{pd_{eq}} \tag{5.18}$$

where

$\mu$ = Absolute viscosity of water (N sec/m² or lb sec/ft²).
$p$ = Density of water (kg/m³ or lb/ft³).
$d_{eq}$ = Here, $d_{90}$ sieve size is used instead of $d_{eq}$.
$G_n$ = Galileo number:

$$\frac{d_{eq}^3 p(p_s - p)g}{\mu^2} \tag{5.19}$$

***Note:*** Based on the studies of Cleasby and Fan (1981), we use a safety factor of 1.3 to ensure adequate movement of the grains.

## ■ Example 5.34

*Problem:* Estimate the minimum fluidized velocity and backwash rate for a sand filter. The $d_{90}$ size of sand is 0.90 mm. The density of the sand is 2.68 g/cm³.

*Solution:* Compute the Galileo number. From the data given and the applicable table, at 15°C:

$p$ = 0.999 g/cm³
$\mu$ = 0.0113 N s/m² = 0.00113 kg/ms = 0.0113 g/cm
$\mu p$ = 0.0113 cm²/s
$g$ = 981 cm/s²
$d$ = 0.90 cm
$p_s$ = 2.68 g/cm³

Using Equation 5.19:

$$\text{Galileo number} = \frac{d_{eq}^3 p(p_s - p)g}{\mu^2} = \frac{(0.090)^3 \times 0.999 \times (2.68 - 0.999) \times 981}{(0.013)^2} = 9405$$

Now compute $U_{mf}$ using Equation 5.19:

$$U_{mf} = \frac{0.0113}{0.999 \times 0.090} \times (1135.69 + 0.0408 \times 9405)^{0.5} - \frac{33.7 \times 0.0113}{0.999 \times 0.090} = 0.660 \text{ cm/sec}$$

Compute the backwash rate. Apply a safety factor of 1.3 to $U_{mf}$ as the backwash rate:

$$\text{Backwash rate} = 1.3 \times 0.660 \text{ cm/sec} = 0.858 \text{ cm/sec}$$

$$0.858 \times \left(\frac{\text{cm}^3}{\text{cm}^2/\text{sec}}\right) \times \frac{1 \text{ L}}{1000 \text{ cm}^3} \times \frac{1}{3.785} \times \frac{\text{gal}}{\text{L}} \times 929 \times \frac{\text{cm}^2}{\text{ft}^2} \times \frac{60 \text{ sec}}{\text{min}} = 12.6 \text{ pg/ft}^2$$

## HORIZONTAL WASTEWATER TROUGHS

Wastewater troughs are used to collect backwash water as well as to distribute influent water during the initial stages of filtration. Troughs are usually placed above the filter media, particularly in the United States. Proper placement of these troughs is very important to ensure that the filter media is not carried into the troughs during the backwash and removed from the filter. These backwash troughs are constructed from concrete, plastic, fiberglass, or other corrosion-resistant materials. The total rate of discharge in a rectangular trough with free flow can be calculated by using Equation 5.20:

$$Q = C \times w \times h^{1.5} \tag{5.20}$$

where

$Q$ = Flow rate (cfs).
$C$ = Constant = 2.49.
$w$ = Trough width (ft).
$h$ = Maximum water depth in trough (ft).

### ■ Example 5.35

*Problem:* Troughs are 18 ft long, 18 in. wide, and 8 ft to the center with a horizontal flat bottom. The backwash rate is 24 in./min. Estimate (1) the water depth of the troughs with free flow into the gullet, and (2) the distance between the top of the troughs and the 30-in. sand bed. Assume 40% expansion, 6 in. of freeboard in the troughs, and 6 in. of thickness.

*Solution:* Estimate the maximum water depth ($h$) in the trough:

$$\text{Velocity } (V) = 24 \text{ in./min} = 2 \text{ ft/60 sec} = 1/30 \text{ fps}$$

$$A = 18 \text{ ft} \times 8 \text{ ft} = 144 \text{ ft}^2$$

$$Q = V \times A = 1/30 \times 144 \text{ cfs} = 4.8 \text{ cfs}$$

Using Equation 5.20:

$$Q = C \times w \times h^{1.5} = 2.49 \times 1.5 \times h^{1.5}$$

$$h = [Q/(2.49 \times 1.5)]^{2/3} = [4.8/(2.49 \times 1.5)]^{2/3} = 1.18 \text{ ft}$$

Now determine the distance ($y$) between the sand bed surface and the trough top:

Freeboard = 6 in. = 0.5 ft
Thickness = 8 in. = 0.67 ft (bottom of trough)

$$y = (2.5 \text{ ft} \times 0.4) + 1.17 \text{ ft} + 0.5 \text{ ft} + 0.5 \text{ ft} = 3.2 \text{ ft}$$

## FILTER EFFICIENCY

Water treatment filter efficiency is defined as the effective filter rate divided by the operation filtration rate as shown in Equation 5.21:

$$E = \frac{R_e}{R_o} = \frac{\text{UFRV} - \text{UBWU}}{\text{UFRV}} \tag{5.21}$$

where
$E$ = Filter efficiency (%).
$R_e$ = Effective filtration rate (gpm/ft²).
$R_o$ = Operating filtration rate (gpm/ft²).
UFRV = Unit filter run volume (gal/ft²).
UBWV = Unit backwash volume (gal/ft²).

### ■ Example 5.36

*Problem:* A rapid sand filter operates at 3.9 gpm/ft² for 48 hr. Upon completion of the filter run, 300 gal/ft² of backwash water are used. Find the filter efficiency.

*Solution:* Calculate operating filtration rate, $R_o$:

$$R_o = 3.9 \text{ gpm/ft}^2 \times 60 \text{ min/hr} \times 48 \text{ hr} = 11{,}232 \text{ gal/ft}^2$$

Calculate effective filtration rate, $R_e$:

$$R_e = 11{,}232 \text{ gal/ft}^2 - 300 \text{ gal/ft}^2 = 10{,}932 \text{ gal/ft}^2$$

Calculate the filter efficiency ($E$) using Equation 5.21:

$$E = 10{,}932 \div 11{,}232 = 97.3\%$$

## WATER FILTRATION PRACTICE PROBLEMS

*Problem 5.1:* A waterworks treats an average of 2.85 MGD. The water is split equally to each of 6 filters. Each filter basin measures 12 ft wide by 22 ft long and 20 ft deep. Each filter bed measures 8 ft by 16 ft by 14 ft deep.

*Problem 5.1a:* Determine the daily flow to each of the filters in gallons per minute.

*Solution:*

$$\frac{2{,}850{,}000 \text{ gpd}}{6 \text{ filters}} = 475{,}000 \text{ gpd}$$

*Problem 5.1b:* The influent line to filter 5 is closed while the effluent remains open. Using a hook gauge and a stop watch, it is noted that the water level in the filter drops 6 inches in 60 sec. What is the filtration rate in gpm?

*Solution:*

$$\text{Volume} = 12 \text{ ft} \times 22 \text{ ft} \times 0.5 \text{ ft} \times 7.48 \text{ gal/ft}^3 = 987.4 \text{ gal}$$

$$\text{Time} = 60 \text{ sec/(60 sec/min)} = 1 \text{ min}$$

$$\text{Filtration rate} = \frac{987.4 \text{ gal}}{1 \text{ min}} = 987.4 \text{ gal/min}$$

*Problem 5.1c:* What is the filtration rate in gallons per minute per filter per square foot of surface area?

*Solution:*

$$\text{Basin area} = 12 \text{ ft} \times 22 \text{ ft} = 264 \text{ ft}^2$$

$$\text{Filtration rate} = \frac{987.4 \text{ gpm}}{264 \text{ ft}^2} = 3.74 \text{ gpm/ft}^2$$

*Problem 5.1d:* A hook gauge was used to determine the rate of rise in the filter bed during the backwash cycle. The water rose 6 inches in 30 sec. What is the backwash rate in gallons per minute?

*Solution:*

$$\text{Volume} = 8 \text{ ft} \times 16 \text{ ft} \times 0.5 \text{ ft} \times 7.48 \text{ gal/ft}^3 = 478.72 \text{ gal}$$

$$\text{Backwash rate} = \frac{478.72 \text{ gal}}{0.5 \text{ min}} = 957.44 \text{ gpm}$$

*Problem 5.1e:* Calculate the filter backwash rate in gallons per minute per square foot.

*Solution:*

$$\text{Area} = 8 \text{ ft} \times 16 \text{ ft} = 128 \text{ ft}^2$$

$$\text{Backwash rate} = \frac{957.44 \text{ gpm}}{128 \text{ ft}^2} = 7.48 \text{ gpm/ft}^2$$

*Problem 5.1f:* Calculate the gallons of water used to backwash the filter if it was backwashed for 15 minutes.

*Solution:*

$$\text{Backwash volume} = \text{Rate (gpm/ft}^2) \times \text{Time (min)} \times \text{Filter area (ft}^2)$$
$$= 7.48 \text{ gpm/ft}^2 \times 15 \text{ min} \times 128 \text{ ft}^2 = 14{,}361.6 \text{ gal}$$

*Problem 5.1g:* During a filter run of 70 hours, the total volume of water filtered was 3.1 million gallons. Calculate the percent of the product water used for backwashing.

*Solution:*

$$\%\text{ Backwash} = \frac{\text{Backwash water (gal)}}{\text{Filtered water (gal)}} \times 100 = \frac{14{,}361 \text{ gal}}{3{,}100{,}000 \text{ gal}} \times 100 = 0.46\%$$

*Problem 5.2:* A filter basin and its sand bed measure 26 ft by 14 ft. Calculate the sand bed area in square feet.

*Solution:*

$$\text{Sand bed area} = \text{Length (ft)} \times \text{Width (ft)} = 26 \text{ ft} \times 14 \text{ ft} = 364 \text{ ft}^2$$

*Problem 5.3:* A filter basin and its sand bed measure 26 ft by 14 ft. If the water drops 6 in., what was the volume (in gallons) of the drop test?

*Solution:*

$$\frac{6 \text{ in.}}{12 \text{ in./ft}} = 0.5 \text{ ft}$$

$$\text{Volume} = 26 \text{ ft} \times 14 \text{ ft} \times 0.5 \text{ ft} \times 7.48 \text{ gal/ft}^3 = 1361.4 \text{ gal}$$

*Problem 5.4:* A filter drop test was timed. The test times were 66 sec, 71 sec, and 70 sec. What was the average time in minutes?

*Solution:*

$$\text{Average time} = \frac{66 \text{ sec} + 71 \text{ sec} + 70 \text{ sec}}{3} = 69 \text{ sec}$$

$$69 \text{ sec} \times \frac{1 \text{ min}}{60 \text{ sec}} = 1.15 \text{ min}$$

*Problem 5.5:* A filter measures 26 ft by 18 ft. When the influent is closed and the effluent is opened, the water drops 6 inches in 2 minutes. What is the filtration rate in gallons per minute?

*Solution:*

$$\frac{6 \text{ in.}}{12 \text{ in./ft}} = 0.5 \text{ ft}$$

$$\text{Filtration rate} = 26 \text{ ft} \times 18 \text{ ft} \times (0.5 \text{ ft/2 min}) \times 7.48 \text{ gal/ft}^3 = \frac{1750.3 \text{ gal}}{2 \text{ min}} = 875.16 \text{ gpm}$$

*Problem 5.6:* A filter measures 26 ft by 18 ft. When the influent is closed and the effluent is opened, the water drains down 6 inches in 2 minutes. What is the filter loading rate in gallons per minute per square foot?

*Solution:*

$$\text{Volume} = 26 \text{ ft} \times 18 \text{ ft} \times 0.5 \text{ ft} \times 7.48 \text{ gal/ft}^3 = 1750.3 \text{ gal}$$

$$\text{Sand area} = 26 \text{ ft} \times 18 \text{ ft} = 468 \text{ ft}^2$$

$$\text{Flow rate} = \frac{1750 \text{ gal}}{2 \text{ min}} = 875 \text{ gpm}$$

$$\text{Filter loading rate} = \frac{875 \text{ gpm}}{468 \text{ ft}^2} = 1.87 \text{ gpm/ft}^2$$

*Problem 5.7:* A filter measures 28 ft by 14 ft. The influent line is shut and the water drops 2.8 in. per minute. Calculate the rate of filtration in million gallons per day.

*Solution:*

$$\frac{2.8 \text{ in.}}{12 \text{ in./ft}} = 0.233 \text{ ft}$$

$$\text{Filtration rate} = \frac{28 \text{ ft} \times 14 \text{ ft} \times 0.233 \text{ ft} \times 7.48 \text{ gal/ft}^3}{1 \text{ min}} = 683.19 \text{ gal}$$

$$\frac{683.19 \text{ gal}}{\text{min}} \times \frac{1440 \text{ min}}{1 \text{ day}} \times \frac{1 \text{ MG}}{1{,}000{,}000 \text{ gal}} = 0.98 \text{ MGD}$$

*Problem 5.8:* A filter measures 26 ft by 14 ft and has a filter media depth of 36 in. Assuming a backwash rate of 14 gpm/ft$^2$ and 10 minutes of backwash required, how many gallons of water are required for each backwash?

*Solution:*

$$\begin{aligned}\text{Backwash volume} &= \text{Backwash rate} \times \text{Time} \times \text{Filter area} \\ &= 14 \text{ gpm/ft}^2 \times 10 \text{ min} \times 26 \text{ ft} \times 14 \text{ ft} \\ &= 50{,}960 \text{ gal}\end{aligned}$$

*Problem 5.9:* The filter in Problem 5.8 filtered 13.90 MG during the last filter run. Based on the gallons produced and the gallons required to backwash the filter, calculate the percent of the product water used for backwashing.

*Solution:*

$$\%\text{ Backwash} = \frac{50{,}960\text{ gal}}{13{,}900{,}000\text{ gal}} \times 100 = 0.367\%$$

*Problem 5.10:* Calculate the filtration rate (in $\text{gpm/ft}^2$) for a filter with a sand area of 28 ft by 21 ft when the applied flow is 2.35 MGD.

*Solution:*

$$\frac{2.35\text{ MG}}{\text{day}} \times \frac{1\text{ day}}{1440\text{ min}} \times \frac{1{,}000{,}000\text{ gal}}{1\text{ MG}} = 1632\text{ gpm}$$

$$\text{Filtration rate} = \frac{1632\text{ gpm}}{28\text{ ft} \times 21\text{ ft}} = 2.78\text{ gpm/ft}^2$$

*Problem 5.11:* Determine the filtration rate (in $\text{gpm/ft}^2$) for a filter with a surface of 24 ft by 18 ft. With the influent valve closed, the water above the filter dropped 12 inches in 5 minutes.

*Solution:*

$$\text{Volume} = 24\text{ ft} \times 18\text{ ft} \times 1\text{ ft} \times 7.48\text{ gal/ft}^3 = \frac{3231.4\text{ gal}}{5\text{ min}} = 646.3\text{ gpm}$$

$$\text{Filtration rate} = \frac{646.3\text{ gpm}}{24\text{ ft} \times 18\text{ ft}} = 1.5\text{ gpm/ft}^2$$

*Problem 5.12:* A filter measures 25 ft by 12 ft. The influent line is shut and the water drops 2.4 inches per minute. Calculate the rate of filtration in MGD.

*Solution:*

$$\frac{2.4\text{ in.}}{12\text{ in./ft}} = 0.2\text{ ft}$$

$$\text{Volume} = 25\text{ ft} \times 12\text{ ft} \times 0.2\text{ ft} \times 7.48\text{ gal/ft}^3 = 448.8\text{ gpm}$$

$$\frac{448.8\text{ gal}}{\text{min}} \times \frac{1440\text{ min}}{\text{day}} \times \frac{1\text{ MG}}{1{,}000{,}000\text{ gal}} = 0.65\text{ MGD}$$

*Problem 5.13:* The filter in Problem 5.12 has a filter media depth of 36 in. Assuming a backwash rate of 14 $\text{gpm/ft}^2$ and 6 minutes of backwash, how many gallons of water are required for each backwash?

*Solution:*

$$\text{Backwash volume} = 14\ \text{gpm/ft}^2 \times 6\ \text{min} \times 25\ \text{ft} \times 12\ \text{ft} = 25{,}200\ \text{gal}$$

*Problem 5.14:* A filter plant has six filters, each measuring 22 ft by 16 ft. One filter is out of commission. The other five filters are capable of filtering 600 gpm. How many gallons per minute per square foot will each filter?

*Solution:*

$$\text{Filtration rate} = \frac{600\ \text{gpm}}{22\ \text{ft} \times 16\ \text{ft}} = 1.7\ \text{gpm/ft}^2$$

*Problem 5.15:* A filter is 30 ft by 18 ft. If, when the influent valve is closed, the water above the filter drops 3.8 inches per minute, what is the rate of filtration in MGD?

*Solution:*

$$\frac{3.8\ \text{in.}}{12\ \text{in./ft}} = 0.317\ \text{ft}$$

$$\text{Filtration rate} = \frac{30\ \text{ft} \times 18\ \text{ft} \times 0.317\ \text{ft} \times 7.48\ \text{gal/ft}^3}{1\ \text{min}} = 1280.4\ \text{gpm}$$

$$\frac{1280.4\ \text{gal}}{\text{min}} \times \frac{1440\ \text{min}}{\text{day}} \times \frac{1\ \text{MG}}{1{,}000{,}000\ \text{gal}} = 1.84\ \text{MGD}$$

*Problem 5.16:* Determine the backwash pumping rate (in gpm) for a filter 32 ft by 20 ft if the desired backwash rate is 16 gpm/ft$^2$.

*Solution:*

$$\text{Backwash pumping rate} = 16\ \text{gpm/ft}^2 \times 32\ \text{ft} \times 20\ \text{ft} = 10{,}240\ \text{gpm}$$

*Problem 5.17:* Determine the volume of water (in gallons) required to backwash the filter in the previous problem if the filter is backwashed for 5 minutes.

*Solution:*

$$\text{Backwash volume} = 16\ \text{gpm/ft}^2 \times 5\ \text{min} \times 32\ \text{ft} \times 20\ \text{ft} = 51{,}200\ \text{gal}$$

*Problem 5.18:* During a filter run the total volume of water filtered was 14.55 million gallons. When the filter was backwashed, 72,230 gallons of water were used. Calculate the percent of the filtered water used for backwashing.

*Solution:*

$$\%\ \text{Backwash} = \frac{72{,}230\ \text{gal}}{14{,}550{,}000\ \text{gal}} \times 100 = 0.5\%$$

## REFERENCES AND RECOMMENDED READING

American Water Works Association and American Society of Civil Engineers. (2004). *Water Treatment Plant Design*, 4th ed. New York: McGraw-Hill.

Chase, G.L. (2002). *Solids Notes: Fluidization*. Akron, OH: The University of Akron.

Cleasby, J.L. (1990). Filtration. In: *Water Quality and Treatment*, American Water Works Association, Eds. New York: McGraw-Hill.

Cleasby, J.L. and Fan, K.S. (1981). Predicting fluidization and expansion of filter media. *J. Environ. Eng.*, 107(EE3): 355–471.

Downs, A.J. and Adams, C.J. (1973). *The Chemistry of Chlorine, Bromine, Iodine and Astatine*. Oxford, UK: Pergamon.

Droste, R.L. (1997). *Theory and Practice of Water and Wastewater Treatment*. New York: John Wiley & Sons.

Fair, G.M., Geyer, J.C., and Okun, D.A. (1968). *Water and Wastewater Engineering*. Vol. 2. *Water Purification and Wastewater Treatment and Disposal*. New York: John Wiley & Sons.

Fetter, C.W. (1998). *Handbook of Chlorination*. New York: Litton Educational.

Gregory, R. and Zabel, T.R. (1990). Sedimentation and flotation. In: *Water Quality and Treatment: A Handbook of Community Water Supplies*, 4th ed., American Water Works Association, Ed. New York: McGraw-Hill.

Gupta, R.S. (1997). *Environmental Engineering and Science: An Introduction*. Rockville, MD: Government Institutes.

Hudson, Jr., H.E. (1989). Density considerations in sedimentation. *J. Am. Water Works Assoc.*, 64(6): 382–386.

McGhee, T.J. (1991). *Water Resources and Environmental Engineering*, 6th ed. New York: McGraw-Hill.

Morris, J.C. (1966). The acid ionization constant of HOCl from 5°C to 35°C. *J. Phys. Chem.*, 70(12): 3789.

Rose, H.E. (1951). On the resistance coefficient—Reynolds number relationship for fluid flow through a bed of granular material. *Proc. Inst. Mech. Eng. (Lond.)*, 153(1): 154–168.

USEPA. (1999). *EPA Guidance Manual Turbidity Provisions*. Washington, DC: U.S. Environmental Protection Agency.

Water, G.C. (1978). *Disinfection of Wastewater and Water for Reuse*. New York: Van Nostrand Reinhold.

Wen, C.Y. and Yu, Y.H. (1966). Minimum fluidization velocity. *AIChE J.*, 12(3): 610–612.

White, G.C. (1972). *Handbook of Chlorination*. New York: Litton.

# 6 Water Chlorination Calculations

Chlorine is the most commonly used substance for disinfection of water in the United States. The addition of chlorine or chlorine compounds to water is called *chlorination*. Chlorination is considered to be the single most important process for preventing the spread of waterborne disease.

## CHLORINE DISINFECTION

Chlorine deactivates microorganisms through several mechanisms that can destroy most biological contaminants:

- It causes damage to the cell wall.
- It alters the permeability of the cell (the ability to pass water in and out through the cell wall).
- It alters the cell protoplasm.
- It inhibits the enzyme activity of the cell so it is unable to use its food to produce energy.
- It inhibits cell reproduction.

Chlorine is available in a number of different forms: (1) as pure elemental gaseous chlorine (a greenish-yellow gas with a pungent and irritating odor), which is heavier than air, nonflammable, nonexplosive, and, when released to the atmosphere, toxic and corrosive; (2) as solid calcium hypochlorite (HTH), in tablets or granules; or (3) as a liquid sodium hypochlorite solution of various strengths. The advantage of one form of chlorine over the others for a given water system depends on the amount of water to be treated, the configuration of the water system, the local availability of the chemicals, and the skill of the operator. One of the major advantages of using chlorine is the effective residual that it produces. A residual indicates that disinfection is completed, and the system has an acceptable bacteriological quality. Maintaining a residual in the distribution system helps to prevent regrowth of microorganisms that were injured but not killed during the initial disinfection stage.

## DETERMINING CHLORINE DOSAGE (FEED RATE)

The units of milligrams per liter (mg/L) and pounds per day (lb/day) are most often used to describe the amount of chlorine added or required. Equation 6.1 can be used to calculate either mg/L or lb/day chlorine dosage.

$$\text{Chlorine feed rate (lb/day)} = \text{Chlorine (mg/L)} \times \text{Flow (MGD)} \times 8.34\text{ lb/gal} \quad (6.1)$$

### ■ Example 6.1

*Problem:* Determine the chlorinator setting (lb/day) required to treat a flow of 4 million gallons per day (MGD) with a chlorine dose of 5 mg/L.

*Solution:*

$$\text{Chlorine feed rate} = \text{Chlorine (mg/L)} \times \text{Flow (MGD)} \times 8.34\ \text{lb/gal}$$

$$= 5\ \text{mg/L} \times 4\ \text{MGD} \times 8.34\ \text{lb/gal} = 167\ \text{lb/day}$$

### ■ Example 6.2

*Problem:* A pipeline 12 inches in diameter and 1400 ft long is to be treated with a chlorine dose of 48 mg/L. How many pounds of chlorine will this require?

*Solution:* Determine the gallon volume of the pipeline:

$$\text{Volume} = 0.785 \times D^2 \times \text{Length (ft)} \times 7.48\ \text{gal/ft}^3$$

$$= 0.785 \times (1\ \text{ft})^2 \times 1400\ \text{ft} \times 7.48\ \text{gal/ft}^3 = 8221\ \text{gal}$$

Now calculate the pounds of chlorine required:

$$\text{Chlorine} = \text{Chlorine (mg/L)} \times \text{Volume (MG)} \times 8.34\ \text{lb/gal}$$

$$= 48\ \text{mg/L} \times 0.008221\ \text{MG} \times 8.34\ \text{lb/gal} = 3.3\ \text{lb}$$

### ■ Example 6.3

*Problem:* A chlorinator setting is 30 lb per 24 hr. If the flow being chlorinated is 1.25 MGD, what is the chlorine dosage expressed as mg/L?

*Solution:*

$$\text{Chlorine} = \text{Chlorine (mg/L)} \times \text{Flow (MGD)} \times 8.34\ \text{lb/gal}$$

$$30\ \text{lb/day} = x\ \text{mg/L} \times 1.25\ \text{MGD} \times 8.34\ \text{lb/gal}$$

$$x = \frac{30\ \text{lb/day}}{1.25\ \text{MGD} \times 8.34\ \text{lb/gal}} = 2.9\ \text{mg/L}$$

### ■ Example 6.4

*Problem:* A flow of 1600 gallons per minute (gpm) is to be chlorinated. At a chlorinator setting of 48 lb per 24 hr, what would be the chlorine dosage in mg/L?

*Solution:* Convert the gpm flow rate to the MGD flow rate:

$$1600\ \text{gpm} \times 1440\ \text{min/day} = 2{,}304{,}000\ \text{gpd} = 2.304\ \text{MGD}$$

Now calculate the chlorine dosage in mg/L:

$$\text{Chlorine} = \text{Chlorine (mg/L)} \times \text{Flow (MGD)} \times 8.34 \text{ lb/gal}$$

$$48 \text{ lb/day} = x \text{ mg/L} \times 2.304 \text{ MGD} \times 8.34 \text{ lb/gal}$$

$$x = \frac{48 \text{ lb/day}}{2.304 \text{ MGD} \times 8.34 \text{ lb/gal}} = 2.5 \text{ mg/L}$$

## CALCULATING CHLORINE DOSE, DEMAND, AND RESIDUAL

Common terms used in chlorination include the following:

- *Chlorine dose*—The amount of chlorine added to the system. It can be determined by adding the desired residual for the finished water to the chlorine demand of the untreated water. Dosage can be either milligrams per liter (mg/L) or pounds per day (lb/day). The most common is mg/L.

$$\text{Chlorine dose (mg/L)} = \text{Chlorine demand (mg/L)} + \text{Chlorine residual (mg/L)}$$

- *Chlorine demand*—The amount of chlorine used by iron, manganese, turbidity, algae, and microorganisms in the water. Because the reaction between chlorine and microorganisms is not instantaneous, demand is relative to time. For example, the demand 5 minutes after applying chlorine will be less than the demand after 20 minutes. Demand, like dosage, is expressed in mg/L. The chlorine demand is as follows:

$$\text{Chlorine demand (mg/L)} = \text{Chlorine dose (mg/L)} - \text{Chlorine residual (mg/L)} \quad (6.2)$$

- *Chlorine residual*—The amount of chlorine (determined by testing) remaining after the demand is satisfied. Residual, like demand, is based on time. The longer the time after dosage, the lower the residual will be, until all of the demand has been satisfied. Residual, like dosage and demand, is expressed in mg/L. The presence of a *free residual* of at least 0.2 to 0.4 ppm usually provides a high degree of assurance that disinfection of the water is complete. *Combined residual* is the result of combining free chlorine with nitrogen compounds; combined residuals are also called chloramines. *Total chlorine residual* is the mathematical combination of free and combined residuals. Total residual can be determined directly with standard chlorine residual test kits.

The following examples illustrate the calculation of chlorine dose, demand, and residual.

### ■ EXAMPLE 6.5

*Problem:* A water sample is tested and found to have a chlorine demand of 1.7 mg/L. If the desired chlorine residual is 0.9 mg/L, what is the desired chlorine dose in mg/L?

*Solution:*

$$\text{Chlorine dose} = \text{Chlorine demand (mg/L)} + \text{Chlorine residual (mg/L)}$$

$$= 1.7 \text{ mg/L} + 0.9 \text{ mg/L} = 2.6 \text{ mg/L}$$

### ■ Example 6.6

*Problem:* The chlorine dosage for water is 2.7 mg/L. If the chlorine residual after a 30-minute contact time is found to be 0.7 mg/L, what is the chlorine demand expressed in mg/L?

*Solution:*

$$\text{Chlorine dose} = \text{Chlorine demand (mg/L)} + \text{Chlorine residual (mg/L)}$$

$$2.7 \text{ mg/L} = x \text{ mg/L} + 0.7 \text{ mg/L}$$

$$2.7 \text{ mg/L} - 0.7 \text{ mg/L} = x \text{ mg/L}$$

$$2.0 \text{ mg/L} = x$$

### ■ Example 6.7

*Problem:* What should the chlorinator setting be (lb/day) to treat a flow of 2.35 MGD if the chlorine demand is 3.2 mg/L and a chlorine residual of 0.9 mg/L is desired?

*Solution:* Determine the chlorine dosage in mg/L:

$$\text{Chlorine dose} = \text{Chlorine demand (mg/L)} + \text{Chlorine residual (mg/L)}$$

$$= 3.2 \text{ mg/L} + 0.9 \text{ mg/L} = 4.1 \text{ mg/L}$$

Calculate the chlorine dosage in lb/day:

$$\text{Chlorine dose} = \text{Chlorine (mg/L)} \times \text{Flow (MGD)} \times 8.34 \text{ lb/gal}$$

$$= 4.1 \text{ mg/L} \times 2.35 \text{ MGD} \times 8.34 \text{ lb/gal} = 80.4 \text{ lb/day}$$

## BREAKPOINT CHLORINATION CALCULATIONS

To produce a free chlorine residual, enough chlorine must be added to the water to produce what is referred to as *breakpoint chlorination*, the point at which nearly complete oxidation of nitrogen compounds is reached; any residual beyond breakpoint is mostly free chlorine (see Figure 6.1). When chlorine is added to natural waters, the chlorine begins combining with and oxidizing the chemicals in the water before it begins disinfecting. Although residual chlorine will be detectable in the water, the chlorine will be in the combined form with a weak disinfecting power. As we see

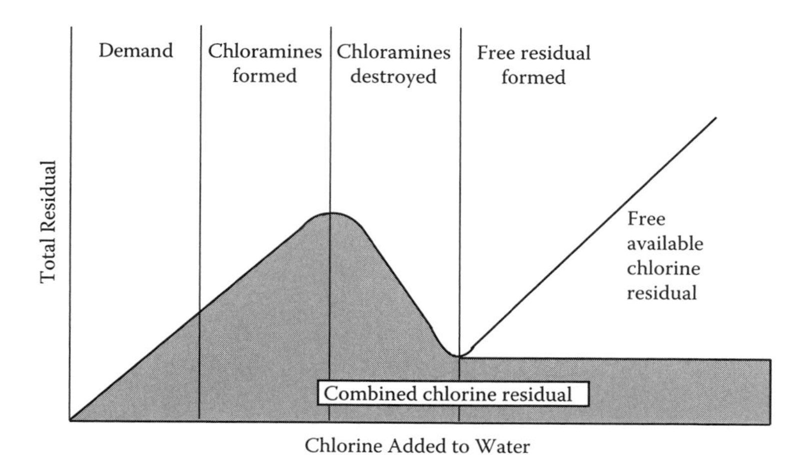


**FIGURE 6.1** Breakpoint chlorination curve.

in Figure 6.1, adding more chlorine to the water at this point actually decreases the chlorine residual as the additional chlorine destroys the combined chlorine compounds. At this stage, water may have a strong swimming pool or medicinal taste and odor. To avoid this, add still more chlorine to produce a free residual chlorine. Free chlorine has the highest disinfecting power. The point at which most of the combined chlorine compounds have been destroyed and the free chlorine starts to form is the *breakpoint.*

***Note:*** The actual chlorine breakpoint of water can only be determined by experimentation.

To calculate the actual increase in chlorine residual that would result from an increase in chlorine dose, we use the mg/L to lb/day equation shown below:

$$\text{Dose increase (lb/day)} = \text{Expected increase (mg/L)} \times \text{Flow (MGD)} \times 8.34 \text{ lb/gal} \quad (6.3)$$

***Note:*** The actual increase in residual is simply a comparison of the new and old residual data.

## ■ Example 6.8

*Problem:* A chlorinator setting is increased by 2 lb/day. The chlorine residual before the increased dosage was 0.2 mg/L. After the increased chlorine dose, the chlorine residual was 0.5 mg/L. The average flow rate being chlorinated is 1.25 MGD. Is the water being chlorinated beyond the breakpoint?

*Solution:* Calculate the expected increase in chlorine residual. Use the mg/L to lb/day equation:

$$\text{Chlorine (lb/day increase)} = \text{Increase (mg/L)} \times \text{Flow (MGD)} \times 8.34 \text{ lb/gal}$$

$$2 \text{ lb/day} = x \text{ mg/L} \times 1.25 \text{ MGD} \times 8.34 \text{ lb/gal}$$

$$x = \frac{2 \text{ lb/day}}{1.25 \text{ MGD} \times 8.34 \text{ lb/gal}} = 0.19 \text{ mg/L}$$

Actual increase in residual is

$$0.5 \text{ mg/L} - 0.19 \text{ mg/L} = 0.31 \text{ mg/L}$$

### ■ Example 6.9

*Problem:* A chlorinator setting of 18 lb chlorine per 24 hr results in a chlorine residual of 0.3 mg/L. The chlorinator setting is increased to 22 lb per 24 hr. The chlorine residual increased to 0.4 mg/L at this new dosage rate. The average flow being treated is 1.4 MGD. On the basis of this information, is the water being chlorinated past the breakpoint?

*Solution:* Calculate the expected increase in chlorine residual:

$$\text{Chlorine (lb/day increase)} = \text{Increase (mg/L)} \times \text{Flow (MGD)} \times 8.34 \text{ lb/gal}$$

$$4 \text{ lb/day} = x \text{ mg/L} \times 1.4 \text{ MGD} \times 8.34 \text{ lb/gal}$$

$$x = \frac{4 \text{ lb/day}}{1.4 \text{ MGD} \times 8.34 \text{ lb/gal}} = 0.34 \text{ mg/L}$$

Actual increase in residual is

$$0.4 \text{ mg/L} - 0.3 \text{ mg/L} = 0.1 \text{ mg/L}$$

## CALCULATING DRY HYPOCHLORITE FEED RATE

The most commonly used dry hypochlorite is calcium hypochlorite (HTH), which contains about 65 to 70% available chlorine, depending on the brand. Because hypochlorites are not 100% pure chorine, more pounds per day must be fed into the system to obtain the same amount of chlorine for disinfection. The equation used to calculate the lb/day hypochlorite required is

$$\text{Hypochlorite (lb/day)} = \frac{\text{Chlorine (lb/day)}}{\text{\% Available chlorine}/100} \tag{6.4}$$

## ■ EXAMPLE 6.10

*Problem:* A chlorine dosage of 110 lb/day is required to disinfect a flow of 1,550,000 gpd. If the calcium hypochlorite to be used contains 65% available chlorine, how many pounds per day hypochlorite will be required for disinfection?

*Solution:* Because only 65% of the hypochlorite is chlorine, more than 110 lb of hypochlorite will be required:

$$\text{Hypochlorite} = \frac{\text{Chlorine (lb/day)}}{\text{\% Available chlorine}/100} = \frac{110 \text{ lb/day}}{65/100} = \frac{110}{0.65} = 169 \text{ lb/day}$$

## ■ EXAMPLE 6.11

*Problem:* A water flow of 900,000 gpd requires a chlorine dose of 3.1 mg/L. If calcium hypochlorite (65% available chlorine) is to be used, how many pounds per day of hypochlorite are required?

*Solution:* Calculate the lb/day chlorine required:

$$\text{Chlorine (lb/day)} = \text{Chlorine (mg/L)} \times \text{Flow (MGD)} \times 8.34 \text{ lb/gal}$$

$$= 3.1 \text{ mg/L} \times 0.90 \text{ MGD} \times 8.34 \text{ lb/gal} = 23 \text{ lb/day}$$

The lb/day hypochlorite can be calculated as

$$\text{Hypochlorite} = \frac{\text{Chlorine (lb/day)}}{\text{\% Available chlorine}/100} = \frac{23 \text{ lb/day}}{65/100} = \frac{23}{0.65} = 35 \text{ lb/day}$$

## ■ EXAMPLE 6.12

*Problem:* A tank contains 550,000 gal of water and is to receive a chlorine dose of 2.0 mg/L. How many pounds of calcium hypochlorite (65% available chlorine) will be required?

*Solution:*

$$\text{Hypochlorite} = \frac{\text{Chlorine (mg/L)} \times \text{Volume (MG)} \times 8.34 \text{ lb/gal}}{\text{\% Available chlorine}/100}$$

$$= \frac{2 \text{ mg/L} \times 0.550 \text{ MG} \times 8.34 \text{ lb/gal}}{65/100}$$

$$= \frac{92}{0.65} = 14.1 \text{ lb/day}$$

### ■ Example 6.13

*Problem:* A total of 40 lb of calcium hypochlorite (65% available chlorine) is used in a day. If the flow rate treated is 1,100,000 gpd, what is the chlorine dosage in mg/L?

*Solution:* Calculate the lb/day chlorine dosage:

$$\text{Hypochlorite (lb/day)} = \frac{\text{Chlorine (lb/day)}}{\text{\% Available chlorine}/100}$$

$$40 \text{ lb/day} = \frac{x \text{ lb/day}}{65/100}$$

$$40 \text{ lb/day} \times 0.65 = x \text{ lb/day}$$

$$26 \text{ lb/day} = x$$

Then calculate mg/L chlorine:

$$26 \text{ lb/day} = x \text{ mg/L} \times 1.10 \text{ MGD} \times 8.34 \text{ lb/gal}$$

$$x \text{ mg/L} = \frac{26 \text{ lb/day}}{1.10 \text{ MGD} \times 8.34 \text{ lb/gal}}$$

$$x = 2.8 \text{ mg/L}$$

### ■ Example 6.14

*Problem:* A flow of 2,550,000 gpd is disinfected with calcium hypochlorite (65% available chlorine). If 50 lb of hypochlorite are used in a 24-hour period, what is the mg/L chlorine dosage?

*Solution:*

$$50 \text{ lb/day hypochlorite} = \frac{x \text{ lb/day chlorine}}{0.65}$$

$$x = 32.5 \text{ lb/day}$$

Calculate the mg/L chlorine:

$$x \text{ mg/L chlorine} \times 2.55 \text{ MGD} \times 8.34 \text{ lb/gal} = 32.5 \text{ lb/day}$$

$$x \text{ mg/L} = \frac{32.5 \text{ lb/day}}{2.55 \text{ MGD} \times 8.34 \text{ lb/gal}}$$

$$x = 1.5 \text{ mg/L}$$

## CALCULATING HYPOCHLORITE SOLUTION FEED RATE

Liquid hypochlorite (i.e., sodium hypochlorite) is supplied as a clear, greenish-yellow liquid in strengths ranging from 5.25% to 16% available chlorine. Often referred to as "bleach," it is, in fact, used for bleaching. Common household bleach is a solution of sodium hypochlorite containing 5.25% available chlorine. When calculating gallons per day liquid hypochlorite, the lb/day hypochlorite required must be converted to gpd hypochlorite required. This conversion is accomplished using Equation 6.5:

$$\text{Hypochlorite (gpd)} = \frac{\text{Hypochlorite (lb/day)}}{8.34\text{ lb/gal}} \tag{6.5}$$

### ■ Example 6.15

*Problem:* A total of 50 lb/day sodium hypochlorite is required for disinfection of a 1.5-MGD flow. How many gallons per day hypochlorite is this?

*Solution:* Because lb/day hypochlorite has already has been calculated, we simply convert lb/day to gpd hypochlorite required:

$$\text{Hypochlorite (gpd)} = \frac{\text{Hypochlorite (lb/day)}}{8.34\text{ lb/gal}} = \frac{50\text{ lb/day}}{8.34\text{ lb/gal}} = 60\text{ gpd}$$

### ■ Example 6.16

*Problem:* A hypochlorinator is used to disinfect the water pumped from a well. The hypochlorite solution contains 3% available chlorine. A chlorine dose of 1.3 mg/L is required for adequate disinfection throughout the system. If the flow being treated is 0.5 MGD, how many gpd of the hypochlorite solution will be required?

*Solution:* Calculate the lb/day chlorine required:

$$1.3\text{ mg/L} \times 0.5\text{ MGD} \times 8.34\text{ lb/gal} = 5.4\text{ lb/day}$$

Calculate the lb/day hypochlorite solution required:

$$\text{Hypochlorite} = \frac{5.4\text{ lb/day chlorine}}{0.03} = 180\text{ lb/day}$$

Calculate the gpd hypochlorite solution required:

$$\text{Hypochlorite (gpd)} = \frac{180\text{ lb/day hypochlorite}}{8.34\text{ lb/gal}} = 21.6\text{ gpd}$$

## CALCULATING PERCENT STRENGTH OF SOLUTIONS

If a teaspoon of salt is dropped into a glass of water it gradually disappears. The salt dissolves in the water. A microscopic examination of the water would not show the salt. Only examination at the molecular level, which is not easily done, would show the salt and water molecules intimately mixed. If we taste the liquid, of course, we would know that the salt is there. And we could recover the salt by evaporating the water. In a solution, the molecules of the salt, the *solute*, are homogeneously dispersed among the molecules of water, the *solvent*. This mixture of salt and water is homogeneous on a molecular level. Such a homogeneous mixture is called a *solution*. The composition of a solution can be varied within certain limits. There are three common states of matter—gas, liquid, and solid. In this discussion, of course, we are only concerned, at the moment, with solids (calcium hypochlorite) and liquids (sodium hypochlorite).

## Calculating Percent Strength Using Dry Hypochlorite

When calculating the percent strength of a chlorine solution, we use Equation 6.6:

$$\%\ \text{Chlorine} = \frac{\text{Hypochlorite (lb)} \times (\%\ \text{Available chlorine}/100)}{\text{Water (lb)} + \begin{bmatrix}\text{Hypochlorite (lb)} \\ \times(\%\ \text{Available chlorine}/100)\end{bmatrix}} \times 100 \tag{6.6}$$

### ■ Example 6.17

*Problem:* If 72 ounces (oz.) of calcium hypochlorite (65% available chlorine) are added to 15 gal of water, what is the percent chlorine strength (by weight) of the solution?

*Solution:*

$$72\ \text{oz.} \div 16\ \text{oz./lb} = 4.5\ \text{lb}$$

$$\%\ \text{Chlorine} = \frac{\text{Hypochlorite (lb)} \times (\%\ \text{Available chlorine}/100)}{\text{Water (lb)} + [\text{Hypochlorite (lb)} \times (\%\ \text{Available chlorine}/100)]} \times 100$$

$$= \frac{4.5\ \text{lb} \times 0.65}{(15\ \text{gal} \times 8.34\ \text{lb/gal}) + (4\ \text{lb} \times 0.65)} \times 100$$

$$= \frac{2.9\ \text{lb}}{125.1\ \text{lb} + 2.6\ \text{lb}} \times 100 = \frac{2.9\ \text{lb}}{127.7\ \text{lb}} \times 100$$

$$= 2.3\%$$

## Calculating Percent Strength Using Liquid Hypochlorite

To calculate percent strength using liquid solutions, such as liquid hypochlorite, a different equation is required:

$$\begin{aligned} &\text{Liquid hypochlorite (gal)} \times 8.34\ \text{lb/gal} \times \frac{\%\ \text{Strength of hypochlorite}}{100} \\ &= \text{Hypochlorite solution (gal)} \times 8.34\ \text{lb/gal} \times \frac{\%\ \text{Strength of hypochlorite}}{100} \end{aligned} \tag{6.7}$$

■ **EXAMPLE 6.18**

*Problem:* A 12% liquid hypochlorite solution is to be used in making up a hypochlorite solution. If 3.3 gal of liquid hypochlorite are mixed with water to produce 25 gal of hypochlorite solution, what is the percent strength of the solution?

*Solution:* Referring to Equation 6.7:

$$3.3 \text{ gal} \times 8.34 \text{ lb/gal} \times \frac{12}{100} = 25 \text{ gal} \times 8.34 \text{ lb/gal} \times \frac{x}{100}$$

$$x = \frac{100 \times 3.3 \text{ gal} \times 8.34 \text{ lb/gal} \times 12}{25 \times 8.34 \text{ lb/gal} \times 100} = 1.6\%$$

## CHEMICAL USE CALCULATIONS

In typical waterworks operation, chemical use (in lb/day or gpd) is documented each day to provide a record of daily use from which the average daily use of the chemical or solution can be calculated. To calculate average use in pounds per day we use Equation 6.8. To calculate average use in gallons per day, we use Equation 6.9.

$$\text{Average use (lb/day)} = \frac{\text{Total chemical used (lb)}}{\text{Number of days}} \tag{6.8}$$

$$\text{Average use (gpd)} = \frac{\text{Total chemical used (gal)}}{\text{Number of days}} \tag{6.9}$$

To calculate the days supply in inventory, we use Equation 6.10 or Equation 6.11.

$$\text{Days supply in inventory} = \frac{\text{Total chemical in inventory (lb)}}{\text{Average use (lb/day)}} \tag{6.10}$$

$$\text{Days supply in inventory} = \frac{\text{Total chemical in inventory (gal)}}{\text{Average use (gpd)}} \tag{6.11}$$

■ **EXAMPLE 6.19**

*Problem:* The pounds of calcium hypochlorite used each day for a week are given below. Based on this information, what was the average lb/day hypochlorite chemical use during the week?

Monday, 50 lb/day
Tuesday, 55 lb/day
Wednesday, 51 lb/day
Thursday, 46 lb/day
Friday, 56 lb/day
Saturday, 51lb/day
Sunday, 48 lb/day

*Solution:*

$$\text{Average use} = \frac{\text{Total chemical used (lb)}}{\text{Number of days}} = \frac{357 \text{ lb}}{7 \text{ days}} = 51 \text{ lb/day}$$

■ **Example 6.20**

*Problem:* The average calcium hypochlorite use at a plant is 40 lb/day. If the chemical inventory in stock is 1100 lb, how many days supply is this?

*Solution:*

$$\text{Days supply in inventory} = \frac{\text{Total chemical in inventory (lb)}}{\text{Average use (lb/day)}} = \frac{1110 \text{ lb}}{40 \text{ lb/day}} = 27.5 \text{ days}$$

## CHLORINATION MATH PRACTICE PROBLEMS

*Problem 6.1:* A waterworks wants to have 1.5 mg/L residual chlorine in the distribution system. Due to a main break, the demand has climbed to 0.9 mg/L. What is the required dose?

*Solution:*

$$\text{Dose} = \text{Demand} + \text{Residual} = 1.5 \text{ mg/L} + 0.9 \text{ mg/L} = 2.4 \text{ mg/L}$$

*Problem 6.2:* A city has a combined residual of 0.6 mg/L and a free residual of 1.6 mg/L. What is the total residual in mg/L?

*Solution:*

$$\text{Total residual} = \text{Combined residual} + \text{Free residual} = 0.6 \text{ mg/L} + 1.6 \text{ mg/L} = 2.2 \text{ mg/L}$$

*Problem 6.3:* A water plant treats 4.1 MGD. If the chlorine dose should be 4.4 mg/L, what is the chlorine feed requirement in lb/day?

*Solution:*

$$\text{Chlorine} = \text{Chlorine (mg/L)} \times \text{Flow (MGD)} \times 8.34 \text{ lb/gal}$$

$$= 4.4 \text{ mg/L} \times 4.1 \text{ MGD} \times 8.34 \text{ lb/gal} = 150.45 \text{ lb/day}$$

*Problem 6.4:* Determine the chlorine dose in mg/L if 16 lb of chlorine are fed while treating 1.2 MGD of water.

*Solution:*

$$\text{Dose} = \frac{\text{Feed rate (lb/day)}}{\text{Flow (MGD)} \times 8.34 \text{ lb/gal}} = \frac{16 \text{ lb}}{1.2 \text{ MGD} \times 8.34 \text{ lb/gal}} = 1.6 \text{ mg/L}$$

*Problem 6.5:* How many pounds of 65% available calcium hypochlorite are required to make 5 gal of an 8% solution?

*Solution:*

$$\text{Calcium hypochlorite} = \frac{(\%\ \text{Solution}/100) \times \text{Volume (gal)} \times 8.34\ \text{lb/gal}}{(\%\ \text{Calcium hypochlorite}/100)}$$

$$= \frac{0.08 \times 5\ \text{gal} \times 8.34\ \text{lb/gal}}{0.65} = 5.13\ \text{lb}$$

*Problem 6.6:* How many gallons of bleach (12.5% available chlorine) will it take to make a 5% solution when added to enough water to make 60 gal of hypochlorite?

*Solution:*

$$\text{Bleach} = \frac{(\%\ \text{Solution}/100) \times \text{Volume (gal)}}{(\%\ \text{Available chlorine}/100)} = \frac{0.05 \times 60\ \text{gal}}{0.125} = 24\ \text{gal}$$

*Problem 6.7:* A waterworks has just switched from sodium hypochlorite (bleach) to chlorine gas. If they used an average of 30 gpd of 15% sodium hypochlorite, how many pounds of $Cl_2$ per day will they use?

*Solution:*

$$\text{Chlorine} = (\%\ \text{Sodium hypochlorite}/100) \times \text{Volume (gal)} \times 8.34\ \text{lb/gal}$$

$$= 0.15 \times 30\ \text{gal} \times 8.34\ \text{lb/gal} = 37.53\ \text{lb}$$

*Problem 6.8:* A water system has a chlorine demand of 4.0 mg/L and wants to have a 1.2 mg/L residual. What would be the dose?

*Solution:*

$$\text{Chlorine dose} - \text{Residual chlorine} = \text{Chlorine demand}$$

$$x\ \text{mg/L} - 1.2\ \text{mg/L} = 4.0\ \text{mg/L}$$

$$x\ \text{mg/L} = 4.0\ \text{mg/L} + 1.2\ \text{mg/L}$$

$$x = 5.2\ \text{mg/L}$$

*Problem 6.9:* A city wants to have 1.5 mg/L chlorine in the distribution system. Due to a main break, the demand has climbed to 1.2 mg/L. What is the residual?

*Solution:*

$$\text{Chlorine dose} - \text{Residual chlorine} = \text{Chlorine demand}$$

$$1.5\ \text{mg/L} - x\ \text{mg/L} = 1.2\ \text{mg/L}$$

$$x\ \text{mg/L} = 1.5\ \text{mg/L} - 1.2\ \text{mg/L}$$

$$x = 0.3\ \text{mg/L}$$

*Problem 6.10:* A system just had a main break. The chlorine level of 3.5 mg/L has dropped to 0.3 mg/L. What is the chlorine demand?

*Solution:*

$$\text{Chlorine demand} = 3.5 \text{ mg/L} - 0.3 \text{ mg/L} = 3.2 \text{ mg/L}$$

*Problem 6.11:* A city doses the water to have a residual of 1.7 mg/L. The demand has risen because of a main break to 1.6 mg/L. What is the free residual?

*Solution:*

$$\begin{aligned} \text{Residual chlorine} &= \text{Chlorine dose} - \text{Chlorine demand} \\ &= 1.7 \text{ mg/L} - 1.6 \text{ mg/L} \\ &= 0.1 \text{ mg/L} \end{aligned}$$

# 7 Fluoridation

The key terms used in this chapter are defined as follows:

- *Fluoride*—The ionic form of fluorine that is found in many waters. It is added to many water systems to reduce tooth decay.
- *Dental caries*—Tooth decay.
- *Dental fluorosis*—Mottled and discolored teeth caused by excessive fluoride content in drinking water.

## WATER FLUORIDATION

As of 2012, 72% of the total U.S. population were getting their water from public water systems that add fluoride. Fluoridation in the United States is being practiced in over 8000 communities serving more than 126 million people. Residents of over 1800 communities serving more than 9 million people are consuming water that contains at least 0.7 mg/L fluoride from natural sources. Key facts about fluoride include the following:

- Fluoride is seldom found in appreciable quantities in surface waters and appears in groundwater in only a few geographical regions.
- Fluoride is sometimes found in a few types of igneous or sedimentary rocks.
- Fluoride is toxic to humans in large quantities; it is also toxic to some animals.
- Based on human experience, fluoride, used in small concentrations (about 1.0 mg/L in drinking water), can be beneficial.

## FLUORIDE COMPOUNDS

Theoretically, any compound that forms fluoride ions in a water solution can be used for adjusting the fluoride content of a water supply; however, there are several practical considerations involved in selecting compounds:

- The compound must have sufficient solubility to permit its use in routine water plant practice.
- The cation to which the fluoride ion is attached must not have any undesirable characteristics.
- The material should be relatively inexpensive and readily available in grades of size and purity suitable for their intended use.

***Caution:*** Fluoride chemicals, like chlorine, caustic soda, and many other chemicals used in water treatment, can constitute a safety hazard for the water plant operator unless proper handling precautions are observed. It is essential that the operator be aware of the hazards associated with each individual chemical prior to its use.

The three commonly used fluoride chemicals should meet the American Water Works Associations (AWWA) standards for use in water fluoridation: sodium fluoride (B701-90), fluorosilicic acid (B703-90), and sodium fluorosilicate (B702-90).

## Sodium Fluoride

The first fluoride compound used in water fluoridation was sodium fluoride. It was selected based on the above criteria and also because its toxicity and physiological effects had been so thoroughly studied. It has become the reference standard for measuring fluoride concentration. Other compounds came into use, but sodium fluoride is still widely used because of its unique physical characteristics. Sodium fluoride (NaF) is a white, odorless material available either as a powder or in the form of crystals of various sizes. It is a salt that in the past was manufactured by adding sulfuric acid to fluorspar and then neutralizing the mixture with sodium carbonate. Neutralizing fluorosilicic acid with caustic soda (NaOH) is now used to produce it. Approximately 19 pounds of sodium fluoride will add 1 ppm of fluoride to 1 million gallons of water. The solubility of sodium fluoride is practically constant at 4 grams per 100 milliliters in water at temperatures generally encountered in water treatment practice (see Table 7.1).

**TABLE 7.1**
**Solubility of Fluoride Chemicals**

| Chemical | Temperature | Solubility (g per 100 mL of $H_2O$) |
|---|---|---|
| Sodium fluoride | 0.0 | 4.00 |
| | 15.0 | 4.03 |
| | 20.0 | 4.05 |
| | 25.0 | 4.10 |
| | 100.0 | 5.00 |
| Sodium fluorosilicate | 0.0 | 0.44 |
| | 25.0 | 0.76 |
| | 37.8 | 0.98 |
| | 65.6 | 1.52 |
| | 100.0 | 2.45 |
| Fluorosilicic acid | Infinite at all temperatures | |

*Source:* Reeves, T.G., *Water Fluoridation: A Manual for Water Plant Operators*, U.S. Department of Health and Human Services, Centers for Disease Control and Prevention, Atlanta, GA, 1994, p. 17.

**TABLE 7.2**
**Properties of Fluorosilicic Acid**

| Acid (%)[a] | Specific Gravity | Density (lb/gal) |
|---|---|---|
| 0 (water) | 1.000 | 8.345 |
| 10 | 1.083 | 9.041 |
| 20 | 1.167 | 9.739 |
| 23 | 1.191 | 9.938 |
| 25 | 1.208 | 10.080 |
| 30 | 1.250 | 10.431 |
| 35 | 1.291 | 10.773 |

*Source:* Reeves, T.G., *Water Fluoridation: A Manual for Water Plant Operators*, U.S. Department of Health and Human Services, Centers for Disease Control and Prevention, Atlanta, GA, 1994, p. 19.

*Note:* Actual densities and specific gravities will be slightly higher when distilled water is not used. Add approximately 0.2 lb/gal to density depending on impurities.

[a] Based on the other percentage being distilled water.

## Fluorosilicic Acid

Fluorosilicic acid ($H_2SiF_6$), also known as hydrofluorosilicic or silicofluoric acid, is a 20 to 35% aqueous solution with a formula weight of 144.08. It is a straw-colored, transparent, fuming, corrosive liquid with a pungent odor and an irritating action on the skin. Solutions of 20 to 35% fluorosilicic acid exhibit a low pH (1.2) and at a concentration of 1 ppm can slightly depress the pH of poorly buffered potable waters. It must be handled with great care because it will cause a delayed burn on skin tissue. The specific gravity and density of fluorosilicic acid are given in Table 7.2.

It takes approximately 46 lb (4.4 gal) of 23% acid to add 1 ppm of fluoride to 1 million gallons of water. Two different processes, resulting in products with differing characteristics, are used to manufacture fluorosilicic acid. The largest production of the acid is as a byproduct of phosphate fertilizer manufacture. Phosphate rock is ground up and treated with sulfuric acid, forming a gas byproduct. Hydrofluoric acid (HF) is an extremely corrosive material. Its presence in fluorosilicic acid, whether from intentional addition (i.e., fortified acid) or from normal production processes demands careful handling of the material.

## Sodium Fluorosilicate

Fluorosilicic acid can readily be converted into various salts, and one of these, sodium fluorosilicate ($Na_2SiF_6$), also known as sodium silicofluoride, is widely used as a chemical for water fluoridation. As with most fluorosilicates, it is generally obtained as a byproduct from the manufacture of phosphate fertilizers. Phosphate rock is ground up and treated with sulfuric acid, forming a gas byproduct. This gas reacts with water and forms fluorosilicic acid. When neutralized with sodium carbonate, sodium fluorosilicate will precipitate out. The conversion of fluorosilicic acid to a dry material containing a high percentage of available fluoride results in a compound that has most of the advantages of the acid with few of its disadvantages. Once it was shown that fluorosilicates form fluoride ions in water solution as readily as do simple fluoride compounds, and that there is no difference in the physiological effect, fluorosilicates were rapidly accepted for water fluoridation and in many cases have displaced the use of sodium fluoride, except in saturators. Sodium fluorosilicate is a white, odorless crystalline powder. Its solubility varies (see Table 7.1). Approximately 14 lb of sodium fluorosilicate will add 1 ppm of fluoride to 1 million gallons of water.

## OPTIMAL FLUORIDE LEVELS

The recommended optimal fluoride concentrations for fluoridated water supply systems are given in Table 7.3. These levels are based on the annual average of the maximum daily air temperature in the area of the involved school or community. In areas where the mean temperature is not shown on Table 7.3, the optimal fluoride level can be determined by the following formula:

$$\text{Fluoride (ppm)} = \frac{0.34}{E} \tag{7.1}$$

where $E$ is the estimated average daily water consumption for children through 10 years of age in ounces of water per pound of body weight. $E$ is obtained from the following formula:

$$E = 0.038 + 0.0062(\text{Average maximum daily air temperature, °F}) \tag{7.2}$$

In Table 7.3, the recommended control range is shifted to the high side of the optimal fluoride level for two reasons:

- It has become obvious that many water plant operators try to maintain the fluoride level in their community at the lowest level possible. The result is that the actual fluoride level in the water will vary around the lowest value in the range instead of around the optimal level.
- Some studies have shown that suboptimal fluorides are relatively ineffective in actually preventing dental caries. Even a drop of 0.2 ppm below optimal levels can reduce dental benefits significantly.

***Note:*** In water fluoridation, underfeeding is a much more serious problem than overfeeding.

**TABLE 7.3**
**Recommended Optimal Fluoride Level**

| Annual Average Maximum Daily Air Temperature (°F)[a] | Recommended Fluoride Concentrations | | Recommended Control Range | | | | | |
|---|---|---|---|---|---|---|---|---|
| | | | Community Systems | | | School Systems | | |
| | Community (ppm) | School (ppm)[b] | 0.1 Below | | 0.5 Above | 20% Low | | 20% High |
| 40.0–53.7 | 1.2 | 5.4 | 1.1 | – | 1.7 | 4.3 | – | 6.5 |
| 53.8–58.3 | 1.1 | 5.0 | 1.0 | – | 1.6 | 4.0 | – | 6.0 |
| 58.4–63.8 | 1.0 | 4.5 | 0.9 | – | 1.5 | 3.6 | – | 5.4 |
| 63.9–70.6 | 0.9 | 4.1 | 0.8 | – | 1.4 | 3.3 | – | 4.9 |
| 70.7–79.2 | 0.8 | 3.6 | 0.7 | – | 1.3 | 2.9 | – | 4.3 |
| 79.3–90.5 | 0.7 | 3.2 | 0.6 | – | 1.2 | 2.6 | – | 3.8 |

*Source:* Reeves, T.G., *Water Fluoridation: A Manual for Water Plant Operators*, U.S. Department of Health and Human Services, Centers for Disease Control and Prevention, Atlanta, GA, 1994, p. 21.

[a] Based on temperature data obtained for a minimum of 5 years.

[b] Based on 4.5 times the optimal fluoride level for communities.

# FLUORIDATION PROCESS CALCULATIONS

## Percent Fluoride Ion in a Compound

When calculating the percent fluoride ion present in a compound, we need to know the chemical formula for the compound (e.g., NaF) and the atomic weight of each element in the compound. The first step is to calculate the molecular weight of each element in the compound (number of atoms × atomic weight = molecular weight). Then we calculate the percent fluoride in the compound using Equation 7.3:

$$\%\text{ Fluoride in compound} = \frac{\text{Molecular weight of fluoride}}{\text{Molecular weight of compound}} \times 100 \tag{7.3}$$

***Note:*** The available fluoride ion concentration is abbreviated as AFI in the calculations that follow. Other important chemical parameters are listed in Table 7.4.

**TABLE 7.4**
**Fluoride Chemical Parameters**

| Chemical | Formula | Available Fluoride Ion (AFI) | Chemical Purity |
|---|---|---|---|
| Sodium fluoride | NaF | 0.453 | 98% |
| Sodium fluorosilicate | $Na_2SiF_6$ | 0.607 | 98% |
| Fluorosilicic acid | $H_2SiF_6$ | 0.792 | 23% |

■ **Example 7.1**

*Problem:* Given the following data, calculate the percent fluoride in sodium fluoride (NaF).

| | No. of Atoms | Molecular Weight |
|---|---|---|
| Na | 1 | 22.997 |
| F | 1 | 19.000 |
| NaF | | 41.997 |

*Solution*: Calculate the percent fluoride in sodium fluoride (NaF):

$$\% \text{ Fluoride in NaF} = \frac{\text{Molecular weight of fluoride}}{\text{Molecular weight of NaF}} \times 100$$

$$= \frac{19.00}{41.997} \times 100$$

$$= 45.2\%$$

## Fluoride Feed Rate

Adjusting the fluoride level in a water supply to an optimal level is accomplished by adding the proper concentration of a fluoride chemical at a consistent rate. To calculate the fluoride feed rate for any fluoridation feeder in terms of pounds of fluoride to be fed per day, it is necessary to determine the following:

- Dosage
- Maximum pumping rate (capacity)
- Chemical purity
- Available fluoride ion concentration

The fluoride feed rate formula is a general equation used to calculate the concentration of a chemical added to water. It will be used for all fluoride chemicals except sodium fluoride when used in a saturator.

***Note:*** mg/L is equal to ppm.

The fluoride feed rate (the amount of chemical required to raise the fluoride content to the optimal level) can be calculated as follows:

$$\text{Fluoride feed rate (lb/day)} = \frac{\text{Dosage (mg/L)} \times \text{Capacity (MGD)} \times 8.34 \text{ lb/gal}}{\text{AFI} \times \text{Chemical purity}} \tag{7.4}$$

If the capacity is in MGD, the fluoride feed rate will be in pounds per day. If the capacity is in gallons per minute (gpm), the feed rate will be pounds per minute if a factor of 1 million is included in the denominator.

$$\text{Fluoride feed rate (lb/min)} = \frac{\text{Dosage (mg/L)} \times \text{Capacity (gpm)} \times 8.34 \text{ lb/gal}}{1{,}000{,}000 \times \text{AFI} \times \text{Chemical purity}} \tag{7.5}$$

## ■ Example 7.2

*Problem:* A water plant produces 2000 gpm and the city wants to add 1.1 mg/L of fluoride. What would the fluoride feed rate be?

*Solution:*

$$2000 \text{ gpm} \times 1440 \text{ min/day} = 2{,}880{,}000 \text{ gpd}$$

$$\frac{2{,}880{,}000 \text{ gpd}}{1{,}000{,}000} = 2.88 \text{ MGD}$$

$$\text{Fluoride feed rate} = \frac{\text{Dosage (mg/L)} \times \text{Capacity (MGD)} \times 8.34 \text{ lb/gal}}{\text{AFI} \times \text{Chemical purity}}$$

$$= \frac{1.1 \text{ mg/L} \times 2.88 \text{ MGD} \times 8.34 \text{ lb/gal}}{0.607 \times 0.985}$$

$$= 44.2 \text{ lb/day}$$

The fluoride feed rate is 44.2 lb/day. Some feed rates from equipment design data sheets are given in grams per minute (g/min). To convert to g/min, divide by 1440 min/day and multiply by 454 g/lb.

$$\text{Fluoride feed rate} = (44.2 \text{ lb/day} \div 1440 \text{ min/day}) \times 454 \text{ g/lb} = 13.9 \text{ g/min}$$

## ■ Example 7.3

*Problem:* If it is known that the plant rate is 4000 gpm and the dosage needed is 0.8 mg/L, what is the fluoride feed rate in mL/min for 23% fluorosilicic acid?

*Solution:*

$$\text{Fluoride feed rate} = \frac{\text{Dosage (mg/L)} \times \text{Capacity (gpm)} \times 8.34 \text{ lb/gal}}{1{,}000{,}000 \times \text{AFI} \times \text{Chemical purity}}$$

$$= \frac{0.8 \text{ mg/L} \times 4000 \text{ gpm} \times 8.34 \text{ lb/gal}}{1{,}000{,}000 \times 0.79 \times 0.23}$$

$$= 0.147 \text{ lb/min}$$

A gallon of 23% fluorosilicic acid weighs 10 lb, and there are 3785 mL per gallon; thus, the following formula can be used to convert the feed rate to mL/min:

$$\text{Feed rate (mL/min)} = (0.147 \text{ lb/min} \div 10 \text{ lb/gal}) \times 3785 \text{ mL/gal} = 55.6 \text{ mL/min}$$

### ■ Example 7.4

*Problem:* If a small water plant wishes to use sodium fluoride in a dry feeder, and the water plant has a capacity (flow) of 180 gpm, what would be the fluoride feed rate? Assume 0.1 mg/L natural fluoride and that 1.0 mg/L is desired in the drinking water.

> ***Note:*** The Centers for Disease Control and Prevention (CDC) recommends against using sodium fluoride in a dry feeder.

*Solution:*

$$\text{Fluoride feed rate} = \frac{\text{Dosage (mg/L)} \times \text{Capacity (gpm)} \times 8.34\text{ lb/gal}}{1{,}000{,}000 \times \text{AFI} \times \text{Chemical purity}}$$

$$= \frac{(1.0 - 0.1\text{ mg/L}) \times 180\text{ gpm} \times 8.34\text{ lb/gal}}{1{,}000{,}000 \times 0.45 \times 0.98}$$

$$= 0.003\text{ lb/min, or } 0.18\text{ lb/hr}$$

Thus, sodium fluoride can be fed at a rate of 0.18 lb/hr to obtain 1.0 mg/L of fluoride in the water.

## Fluoride Feed Rates for Saturator

A sodium fluoride saturator is unique in that the strength of the saturated solution formed is always 18,000 ppm. This is because sodium fluoride has a solubility that is practically constant at 4.0 grams per 100 milliliters of water at temperatures generally encountered in water treatment. This means that each liter of solution contains 18,000 mg of fluoride ion: 40,000 mg/L × percent available fluoride (45%) = 18,000 mg/L. This consistency simplifies calculations because it eliminates the need for weighing the chemicals. A meter on the water inlet of the saturator provides this volume; all that is needed is the volume of solution added to the water to calculate dosage.

$$\text{Fluoride feed rate (gpm)} = \frac{\text{Capacity (gpm)} \times \text{Dosage (mg/L)}}{18{,}000\text{ mg/L}} \tag{7.6}$$

The fluoride feed rate will have the same units as the capacity. If the capacity is in gallons per minute (gpm), the feed rate will also be in gpm. If the capacity is in gallons per day (gpd), the feed rate will also be in gpd.

To change the fluoride feed rate from pounds of dry feed to gallons of solution, divide by the concentration of sodium fluoride and the density of the solution (water).

> ***Note:*** The chemical purity of the sodium fluoride in solution will be 4% × 8.34 lb/gal.

$$\text{Fluoride feed rate (gpm)} = \frac{\text{Dosage (mg/L)} \times \text{Capacity (gpm)} \times 8.34 \text{ lb/gal}}{1{,}000{,}000 \times \text{AFI} \times \text{Chemical purity}} \tag{7.7}$$

$$= \frac{\text{Dosage (mg/L)} \times \text{Capacity (gpm)} \times 8.34 \text{ lb/gal}}{1{,}000{,}000 \times 0.45 \times 4\% \times 8.34 \text{ lb/gal}}$$

$$= \frac{\text{Dosage (mg/L)} \times \text{Capacity (gpm)}}{1{,}000{,}000 \times 0.45 \times 4\%}$$

$$= \frac{\text{Dosage (mg/L)} \times \text{Capacity (gpm)}}{18{,}000 \text{ mg/L}}$$

### ■ Example 7.5

*Problem:* A water plant produces 1 MGD and has less than 0.1 mg/L of natural fluoride. What must the fluoride feed rate be to obtain 1 mg/L in the water?

*Solution:*

$$\text{Fluoride feed rate} = \frac{\text{Dosage (mg/L)} \times \text{Capacity (gpm)}}{18{,}000 \text{ mg/L}}$$

$$= \frac{1 \text{ mg/L} \times 1{,}000{,}000 \text{ gpd}}{18{,}000 \text{ mg/L}}$$

$$= 55.6 \text{ gpd}$$

Thus, it takes approximately 56 gal of saturated solution to treat 1 MG of water at a dose of 1 mg/L.

## Calculated Dosages

Some states require that records be kept regarding the amount of chemical used and that the theoretical concentration of a chemical in the water be determined mathematically. In order to find the theoretical concentration of fluoride, the calculated dosage must be determined. Adding the calculated dosage to the natural fluoride level in the water supply will yield the theoretical concentration of fluoride in the water. This number, the theoretical concentration, is calculated as a safety precaution to help ensure that an overfeed or accident does not occur. It is also an aid in troubleshooting problems. If the theoretical concentration is significantly higher or lower than the measured concentration, steps should be taken to determine the discrepancy. The fluoride feed rate formula can be changed to find the calculated dosage as follows:

$$\text{Dosage (mg/L)} = \frac{\text{Fluoride feed rate (lb/day)} \times \text{AFI} \times \text{Chemical purity}}{\text{Capacity (MGD)} \times 8.34 \text{ lb/gal}} \tag{7.8}$$

When the fluoride feed rate is changed to fluoride fed and the capacity is changed to actual daily production of water in the system, then the dosage becomes the calculated dosage.

> ***Note:*** The amount of fluoride fed (lb) will be determined over a time period (e.g., day, week, month) and the actual production will be determined over the same time period.

$$\text{Calculated dosage (mg/L)} = \frac{\text{Fluoride fed (lb)} \times \text{AFI} \times \text{Chemical purity}}{\text{Actual production (MG)} \times 8.34 \text{ lb/gal}} \tag{7.9}$$

The numerator of the equation gives the pounds of fluoride ion added to the water, and the denominator gives million pounds of water treated. Pounds of fluoride divided by million pounds of water equals ppm or mg/L.

The formula for the calculated dosage for the saturator is as follows:

$$\text{Calculated dosage (mg/L)} = \frac{\text{Solution fed (gal)} \times 18{,}000 \text{ mg/L}}{\text{Actual production (gal)}} \tag{7.10}$$

Determining the calculated dosage for an unsaturated sodium fluoride solution is based on the particular strength of the solution. For example, a 2% strength solution is equal to 8550 mg/L. The percent strength is based on the pounds of sodium fluoride dissolved into a certain amount of water. For example, find the percent solution if 6.5 lb of sodium fluoride are dissolved in 45 gal of water:

$$45 \text{ gal} \times 8.34 \text{ lb/gal} = 375 \text{ lb of water}$$

$$\frac{6.5 \text{ lb sodium fluoride}}{375 \text{ lb water}} = 1.7\% \text{ sodium fluoride solution}$$

This means that 6.5 lb of fluoride chemical dissolved in 45 gal of water will yield a 1.7% solution. To find the solution concentration of an unknown sodium fluoride solution, use the following formula:

$$\text{Solution concentration} = \frac{18{,}000 \text{ mg/L} \times \text{Solution strength (\%)}}{4\%} \tag{7.11}$$

### ■ Example 7.6

*Problem:* Assume that 6.5 lb of sodium fluoride are dissolved in 45 gal of water, as previously given. What would be the solution concentration? Solution strength is 1.7% (see above).

*Solution:*

$$\text{Solution concentration} = \frac{18{,}000 \text{ mg/L} \times \text{Solution strength (\%)}}{4\%}$$

$$= \frac{18{,}000 \text{ mg/L} \times 1.7\%}{4\%} = 7650 \text{ mg/L}$$

The calculated dosage formula for an unsaturated sodium fluoride solution is

$$\text{Calculated dosage (mg/L)} = \frac{\text{Solution fed (gal)} \times \text{Solution concentration (mg/L)}}{\text{Actual production (gal)}}$$

***Note:*** The CDC recommends against the use of unsaturated sodium fluoride solution in water fluoridation.

## Calculated Dosage Examples

### ■ Example 7.7

*Problem:* A plant uses 65 lb of sodium fluorosilicate to treat 5,540,000 gal of water in one day. What is the calculated dosage?

*Solution:*

$$\text{Calculated dosage} = \frac{\text{Fluoride fed (lb)} \times \text{AFI} \times \text{Purity}}{\text{Actual production (MG)} \times 8.34 \text{ lb/gal}}$$

$$= \frac{65 \text{ lb} \times 0.607 \times 0.985}{5.54 \text{ MG} \times 8.34 \text{ lb/gal}} = 0.84 \text{ mg/L}$$

### ■ Example 7.8

*Problem:* A plant uses 43 lb of fluorosilicic acid to treat 1,226,000 gal of water. Assume the purity of the acid is 23%. What is the calculated dosage?

*Solution:*

$$\text{Calculated dosage} = \frac{\text{Fluoride fed (lb)} \times \text{AFI} \times \text{Purity}}{\text{Actual production (MG)} \times 8.34 \text{ lb/gal}}$$

$$= \frac{43 \text{ lb} \times 0.792 \times 0.23}{1.226 \text{ MG} \times 8.34 \text{ lb/gal}} = 0.77 \text{ mg/L}$$

The calculated dosage is 0.77 mg/L. If the natural fluoride level is added to this dosage, then it should equal what the actual fluoride level is in the drinking water.

### ■ Example 7.9

*Problem:* A water plant feeds sodium fluoride in a dry feeder. They use 5.5 lb of the chemical to fluoridate 240,000 gal of water. What is the calculated dosage?

*Solution:*

$$\text{Calculated dosage} = \frac{\text{Fluoride fed (lb)} \times \text{AFI} \times \text{Purity}}{\text{Actual production (MG)} \times 8.34\ \text{lb/gal}}$$

$$= \frac{5.5\ \text{lb} \times 0.45 \times 0.98}{0.24\ \text{MG} \times 8.34\ \text{lb/gal}} = 1.2\ \text{mg/L}$$

### ■ Example 7.10

*Problem:* A plant uses 10 gal of sodium fluoride from its saturator to treat 200,000 gal of water. What is the calculated dosage?

*Solution:*

$$\text{Calculated dosage} = \frac{\text{Solution fed (gal)} \times 18{,}000\ \text{mg/L}}{\text{Actual production (gal)}}$$

$$= \frac{10\ \text{gal} \times 18{,}000\ \text{mg/L}}{200{,}000\ \text{gal}} = 0.9\ \text{mg/L}$$

### ■ Example 7.11

*Problem:* A water plant adds 93 gal per day of a 2% solution of sodium fluoride to fluoridate 800,000 gpd. What is the calculated dosage?

*Solution:*

$$\text{Solution concentration} = \frac{18{,}000\ \text{mg/L} \times \text{Solution strength (\%)}}{4\%}$$

$$= \frac{18{,}000\ \text{mg/L} \times 0.02}{0.04} = 9000\ \text{mg/L}$$

$$\text{Calculated dosage} = \frac{\text{Solution fed (gal)} \times \text{Solution concentration (mg/L)}}{\text{Actual production (gal)}}$$

$$= \frac{93\ \text{gal} \times 9000\ \text{mg/L}}{800{,}000\ \text{gal}} = 1.05\ \text{mg/L}$$

## FLUORIDATION PRACTICE PROBLEMS

For the problems below, the AFI and purity parameters can be found in Table 7.4.

*Problem 7.1:* A water plant produces 1800 gpm and the town wants to have 0.9 mg/L of fluoride in the finished water. If fluorosilicic acid is used, what would be the fluoride feed rate in lb/day?

*Solution:*

$$(1800 \text{ gal/min}) \times (60 \text{ min/day}) \times (1 \text{ MG}/1{,}000{,}000 \text{ gal}) = 0.108 \text{ MGD}$$

$$\text{Feed rate} = \frac{\text{Dose (mg/L)} \times \text{Flow (MGD)} \times 8.34 \text{ lb/gal}}{\text{AFI} \times \text{Purity}}$$

$$= \frac{0.9 \text{ mg/L} \times 0.108 \text{ MGD} \times 8.34 \text{ lb/gal}}{0.792 \times 0.23}$$

$$= 3.85 \text{ lb/day}$$

*Problem 7.2:* A water plant produces 2650 gpm. What would be the fluoride feed rate (in gpm) from a saturator to obtain 0.7 mg/L in the water?

*Solution:*

$$\text{Feed rate} = \frac{\text{Dose (mg/L)} \times \text{Flow (gpm)}}{18{,}000 \text{ mg/L}} = \frac{0.7 \text{ mg/L} \times 2650 \text{ gpm}}{18{,}000 \text{ mg/L}} = 0.10 \text{ gpm}$$

*Problem 7.3:* A plant uses 80 lb of sodium fluorosilicate to treat 9.0 MGD. What is the calculated dosage in mg/L?

*Solution:*

$$\text{Calculated dosage} = \frac{\text{Sodium fluorosilicate (lb)} \times \text{AFI} \times \text{Purity}}{\text{Flow (MGD)} \times 8.34 \text{ lb/gal}}$$

$$= \frac{80 \text{ lb} \times 0.985 \times 0.607}{9 \text{ MGD} \times 8.34 \text{ lb/gal}} = 0.64 \text{ mg/L}$$

*Problem 7.4:* The fluoride for a plant's raw water source was measured to be 0.2 mg/L. If the city wants that finished water to contain the recommended amount of 0.6 mg/L, what mg/L of fluoride should the water plant dose?

*Solution:*

$$\text{Fluoride} = 0.6 \text{ mg/L} - 0.2 \text{ mg/L} = 0.4 \text{ mg/L}$$

*Problem 7.5:* A water plant has a daily average production of 660 gpm, and the city wants to have 1.0 mg/L fluoride in the finished water. The natural fluoride level is less than 0.1 mg/L. Find the fluoride feed rate in lb/day using sodium fluorosilicate.

*Solution:*

$$(660 \text{ gal/min}) \times (1440 \text{ min/day}) \times (1 \text{ MG}/1{,}000{,}000 \text{ gal}) = 0.9504 \text{ MGD}$$

$$\text{Feed rate} = \frac{\text{Dose (mg/L)} \times \text{Flow (MGD)} \times 8.34 \text{ lb/gal}}{\text{AFI} \times \text{Purity}}$$

$$= \frac{1.0 \text{ mg/L} \times 0.9504 \text{ MGD} \times 8.34 \text{ lb/gal}}{0.607 \times 0.985}$$

$$= 13.25 \text{ lb/day}$$

*Problem 7.6:* If it is known that the plant rate is 4200 gpm and the dosage required is 0.9 mg/L, what is the fluoride feed rate (in lb/min) using fluorosilicic acid?

*Solution:*

$$\text{Feed rate} = \frac{\text{Dose (mg/L)} \times \text{Flow (gpm)} \times 8.34 \text{ lb/gal}}{1{,}000{,}000 \times \text{AFI} \times \text{Purity}}$$

$$= \frac{0.9 \text{ mg/L} \times 4200 \text{ gpm} \times 8.34 \text{ lb/gal}}{1{,}000{,}000 \times 0.79 \times 0.23}$$

$$= 0.17 \text{ lb/min}$$

*Problem 7.7:* What is the fluoride feed rate (in lb/day) using fluorosilicic acid if the plant rate is 1.0 MGD, the natural fluoride content is 0.3 mg/L, and the desired fluoride content is 1.3 mg/L?

*Solution:*

$$\text{Dose} = 1.3 \text{ mg/L} - 0.3 \text{ mg/L} = 1.0 \text{ mg/L}$$

$$\text{Feed rate} = \frac{\text{Dose (mg/L)} \times \text{Flow (MGD)} \times 8.34 \text{ lb/gal}}{\text{AFI} \times \text{Purity}}$$

$$= \frac{1.0 \text{ mg/L} \times 1.0 \text{ MGD} \times 8.34 \text{ lb/gal}}{0.792 \times 0.23}$$

$$= 45.8 \text{ lb/day}$$

*Problem 7.8:* If a waterworks wishes to use sodium fluorosilicate in a dry feeder and the water plant has a flow of 210 gpm, what would be the fluoride feed rate in lb/min? Assume 0.1 mg/L natural fluoride and that 1.1 mg/L is the desired concentration in the finished water.

*Solution:*

$$\text{Dose} = 1.1 \text{ mg/L} - 0.1 \text{ mg/L} = 1.0 \text{ mg/L}$$

$$\text{Feed rate} = \frac{\text{Dose (mg/L)} \times \text{Flow (gpm)} \times 8.34 \text{ lb/gal}}{1{,}000{,}000 \times \text{AFI} \times \text{Purity}}$$

$$= \frac{1.0 \text{ mg/L} \times 210 \text{ gpm} \times 8.34 \text{ lb/gal}}{1{,}000{,}000 \times 0.607 \times 0.985}$$

$$= 0.003 \text{ lb/min}$$

*Problem 7.9:* A waterworks produces 1.1 MGD. What would be the fluoride feed rate (in gpd) from a saturator to obtain 1.1 mg/L in the water?

*Solution:*

$$(1.1 \text{ MG/day}) \times (1 \text{ day/1440 min}) \times (1{,}000{,}000 \text{ gal/1 MG}) = 764 \text{ gpm}$$

$$\text{Feed rate} = \frac{\text{Dose (mg/L)} \times \text{Flow (gpm)}}{18{,}000 \text{ mg/L}} = \frac{1.1 \text{ mg/L} \times 764 \text{ gpm}}{18{,}000 \text{ mg/L}} = 0.047 \text{ gpm}$$

$$(0.047 \text{ gal/min}) \times (1440 \text{ min/day}) = 67.7 \text{ gpd}$$

# 8 Water Softening

## WATER HARDNESS

Hardness in water is caused by the presence of certain positively charged metallic ions in solution in the water. The most common of these hardness-causing ions are calcium and magnesium; others include iron, strontium, and barium. The two primary constituents of water that determine the hardness of water are calcium and magnesium. If the concentration of these elements in the water is known, the total hardness of the water can be calculated. To make this calculation, the equivalent weights of calcium, magnesium, and calcium carbonate must be known; the equivalent weights are given below:

| | |
|---|---|
| Calcium (Ca) | 20.04 |
| Magnesium (Mg) | 12.15 |
| Calcium carbonate ($CaCO_3$) | 50.045 |

### Calculating Calcium Hardness (as $CaCO_3$)

The hardness (in mg/L as $CaCO_3$) for any given metallic ion is calculated using Equation 8.1.

$$\frac{\text{Calcium hardness (mg/L as CaCO}_3\text{)}}{\text{Equivalent weight of CaCO}_3} = \frac{\text{Calcium (mg/L)}}{\text{Equivalent weight of calcium}} \tag{8.1}$$

#### ■ Example 8.1

*Problem:* A water sample has calcium content of 51 mg/L. What is this calcium hardness expressed as $CaCO_3$?

*Solution:*

$$\frac{\text{Calcium hardness (mg/L as CaCO}_3\text{)}}{\text{Equivalent weight of CaCO}_3} = \frac{\text{Calcium (mg/L)}}{\text{Equivalent weight of calcium}}$$

$$\frac{x \text{ mg/L as CaCO}_3}{50.045} = \frac{51 \text{ mg/L}}{20.04}$$

$$x \text{ mg/L as CaCO}_3 = \frac{50.045 \times 51 \text{ mg/L}}{20.04}$$

$$x = 124.8 \text{ mg/L as CaCO}_3$$

### ■ Example 8.2

*Problem:* The calcium content of a water sample is 26 mg/L. What is this calcium hardness expressed as $CaCO_3$?

*Solution:*

$$\frac{x \text{ mg/L as CaCO}_3}{50.045} = \frac{26 \text{ mg/L}}{20.04}$$

$$x \text{ mg/L as CaCO}_3 = \frac{50.045 \times 26 \text{ mg/L}}{20.04}$$

$$x = 64.9 \text{ mg/L as CaCO}_3$$

## Calculating Magnesium Hardness (as $CaCO_3$)

To calculate magnesium hardness, we use Equation 8.2.

$$\frac{\text{Magnesium hardness (mg/L as CaCO}_3)}{\text{Equivalent weight of CaCO}_3} = \frac{\text{Magnesium (mg/L)}}{\text{Equivalent weight of magnesium}} \tag{8.2}$$

### ■ Example 8.3

*Problem:* A sample of water contains 24 mg/L magnesium. Express this magnesium hardness as $CaCO_3$.

*Solution:*

$$\frac{\text{Magnesium hardness (mg/L as CaCO}_3)}{\text{Equivalent weight of CaCO}_3} = \frac{\text{Magnesium (mg/L)}}{\text{Equivalent weight of magnesium}}$$

$$\frac{x \text{ mg/L as CaCO}_3}{50.045} = \frac{24 \text{ mg/L}}{12.15}$$

$$x \text{ mg/L as CaCO}_3 = \frac{50.045 \times 24 \text{ mg/L}}{12.15}$$

$$x = 98.9 \text{ mg/L as CaCO}_3$$

### ■ Example 8.4

*Problem:* The magnesium content of a water sample is 16 mg/L. Express this magnesium hardness as $CaCO_3$.

*Solution:*

$$\frac{x \text{ mg/L as CaCO}_3}{50.045} = \frac{16 \text{ mg/L}}{12.15}$$

$$x \text{ mg/L as CaCO}_3 = \frac{50.045 \times 16 \text{ mg/L}}{12.15}$$

$$x = 65.9 \text{ mg/L as CaCO}_3$$

## Calculating Total Hardness

Calcium and magnesium ions are the two constituents that are the primary cause of hardness in water. To find total hardness, we simply add the concentrations of calcium and magnesium ions, expressed in terms of calcium carbonate ($CaCO_3$) using Equation 8.3:

$$\text{Total hardness (mg/L as CaCO}_3) = \text{Calcium hardness (mg/L as CaCO}_3) + \text{Magnesium hardness (mg/L as CaCO}_3) \quad (8.3)$$

### ■ Example 8.5

*Problem:* A sample of water has calcium content of 70 mg/L as $CaCO_3$ and magnesium content of 90 mg/L as $CaCO_3$. What is the total hardness?

*Solution:*

$$\text{Total hardness} = \text{Calcium hardness} + \text{Magnesium hardness}$$

$$= 70 \text{ mg/L} + 90 \text{ mg/L} = 160 \text{ mg/L as CaCO}_3$$

### ■ Example 8.6

*Problem:* Determine the total hardness as $CaCO_3$ of a sample of water that has calcium content of 28 mg/L and magnesium content of 9 mg/L.

*Solution:* Express calcium and magnesium in terms of $CaCO_3$:

Calcium hardness:

$$\frac{x \text{ mg/L as CaCO}_3}{50.045} = \frac{28 \text{ mg/L}}{20.04}$$

$$x \text{ mg/L as CaCO}_3 = \frac{50.045 \times 28 \text{ mg/L}}{20.04}$$

$$x = 69.9 \text{ mg/L as CaCO}_3$$

Magnesium hardness:

$$\frac{x \text{ mg/L as CaCO}_3}{50.045} = \frac{9 \text{ mg/L}}{12.15}$$

$$x \text{ mg/L as CaCO}_3 = \frac{50.045 \times 9 \text{ mg/L}}{12.15}$$

$$x = 37.1 \text{ mg/L as CaCO}_3$$

Now, total hardness can be calculated:

$$\text{Total hardness} = \text{Calcium hardness} + \text{Magnesium hardness}$$

$$= 69.9 \text{ mg/L} + 37.1 \text{ mg/L} = 107 \text{ mg/L as CaCO}_3$$

## Calculating Carbonate and Noncarbonate Hardness

Total hardness is comprised of calcium and magnesium hardness. Once total hardness has been calculated, it is sometimes used to determine another expression of hardness: carbonate and noncarbonate. When hardness is numerically greater than the sum of bicarbonate and carbonate alkalinity, that amount of hardness equivalent to the total alkalinity (both in units of mg/L as $CaCO_3$) is the *carbonate hardness*; the amount of hardness in excess of this is the *noncarbonate hardness*. When the hardness is numerically equal to or less than the sum of carbonate and noncarbonate alkalinity, all hardness is carbonate hardness and noncarbonate hardness is absent.

Again, the total hardness is comprised of carbonate hardness and noncarbonate hardness:

$$\text{Total hardness} = \text{Carbonate hardness} + \text{Noncarbonate hardness} \tag{8.4}$$

When the alkalinity (as $CaCO_3$) is *greater than* the total hardness, all of the hardness is carbonate hardness:

$$\text{Total hardness (mg/L as CaCO}_3\text{)} = \text{Carbonate hardness (mg/L as CaCO}_3\text{)} \tag{8.5}$$

When the alkalinity (as $CaCO_3$) is *less than* the total hardness, then the alkalinity represents carbonate hardness and the balance of the hardness is noncarbonate hardness:

$$\begin{aligned}\text{Total hardness (mg/L as CaCO}_3\text{)} = {} & \text{Carbonate hardness (mg/L as CaCO}_3\text{)} \\ & + \text{Noncarbonate hardness (mg/L as CaCO}_3\text{)}\end{aligned} \tag{8.6}$$

When carbonate hardness is represented by the alkalinity, we use Equation 8.7.

$$\begin{aligned}\text{Total hardness (mg/L as CaCO}_3\text{)} = {} & \text{Alkalinity (mg/L as CaCO}_3\text{)} \\ & + \text{Noncarbonate hardness (mg/L as CaCO}_3\text{)}\end{aligned} \tag{8.7}$$

### ■ EXAMPLE 8.7

*Problem:* A water sample contains 110-mg/L alkalinity as $CaCO_3$ and 105 mg/L total hardness as $CaCO_3$. What are the carbonate hardness and noncarbonate hardness of the sample?

*Solution:* Because the alkalinity is greater than the total hardness, all of the hardness is carbonate hardness:

$$\text{Total hardness} = \text{Carbonate hardness (mg/L as CaCO}_3\text{)} = 105\text{ mg/L as CaCO}_3$$

No noncarbonate hardness is present in this water.

### ■ EXAMPLE 8.8

*Problem:* The alkalinity of a water sample is 80 mg/L as $CaCO_3$. If the total hardness of the water sample is 112 mg/L as $CaCO_3$, what are the carbonate hardness and noncarbonate hardness in mg/L as $CaCO_3$?

*Solution:* Alkalinity is less than total hardness; therefore, both carbonate and noncarbonate hardness will be present in the hardness of the sample.

$$\text{Total hardness (mg/L as CaCO}_3) = \text{Carbonate hardness (mg/L as CaCO}_3) + \text{Noncarbonate hardness (mg/L as CaCO}_3)$$

$$112 \text{ mg/L} = 80 \text{ mg/L} - x \text{ mg/L}$$

$$112 \text{ mg/L} - 80 \text{ mg/L} = x \text{ mg/L}$$

$$32 \text{ mg/L} = x$$

## DETERMINING ALKALINITY

Alkalinity measures the acid-neutralizing capacity of a water sample. It is an aggregate property of the water sample and can be interpreted in terms of specific substances only when a complete chemical composition of the sample is also performed. The alkalinity of surface waters is primarily due to the carbonate, bicarbonate, and hydroxide content and is often interpreted in terms of the concentrations of these constituents. The higher the alkalinity, the greater the capacity of the water to neutralize acids; conversely, the lower the alkalinity, the less the neutralizing capacity. To detect the different types of alkalinity, the water is tested for phenolphthalein and total alkalinity, and Equations 8.8 and 8.9 are used.

$$\text{Phenolphthalein alkalinity (mg/L as CaCO}_3) = \frac{A \times N \times 50{,}000}{\text{Sample (mL)}} \tag{8.8}$$

$$\text{Total alkalinity (mg/L as CaCO}_3) = \frac{B \times N \times 50{,}000}{\text{Sample (mL)}} \tag{8.9}$$

where

$A$ = Titrant (mL) used to titrate to pH 8.3.
$N$ = Normality of the acid (0.02-$N$ $H_2SO_4$ for this alkalinity test).
$B$ = Total titrant (mL) used to titrate to pH 4.5.
50,000 = Conversion factor to change the normality into units of $CaCO_3$.

### ■ EXAMPLE 8.9

*Problem:* A 100-mL water sample is tested for phenolphthalein alkalinity. If 1.3 mL titrant are used to titrate to pH 8.3 and the sulfuric acid solution has a normality of 0.02 $N$, what is the phenolphthalein alkalinity of the water?

*Solution:*

$$\text{Phenolphthalein alkalinity} = \frac{A \times N \times 50{,}000}{\text{Sample (mL)}}$$

$$= \frac{1.3 \text{ mL} \times 0.02\ N \times 50{,}000}{100 \text{ mL}} = 13 \text{ mg/L as CaCO}_3$$

■ **Example 8.10**

*Problem:* A 100-mL sample of water is tested for alkalinity. The normality of the sulfuric acid used for titrating is 0.02 *N*. If 0 mL is used to titrate to pH 8.3, and 7.6 mL titrant is used to titrate to pH 4.5, what are the phenolphthalein alkalinity and total alkalinity of the sample?

*Solution:*

$$\text{Phenolphthalein alkalinity} = \frac{A \times N \times 50{,}000}{\text{Sample (mL)}}$$

$$= \frac{0 \text{ mL} \times 0.02\ N \times 50{,}000}{100 \text{ mL}} = 0 \text{ mg/L as } CaCO_3$$

$$\text{Total alkalinity} = \frac{B \times N \times 50{,}000}{\text{Sample (mL)}}$$

$$= \frac{7.6 \text{ mL} \times 0.02\ N \times 50{,}000}{100 \text{ mL}} = 76 \text{ mg/L as } CaCO_3$$

## DETERMINING BICARBONATE, CARBONATE, AND HYDROXIDE ALKALINITY

Interpretation of phenolphthalein and total alkalinity test results (assuming all of the alkalinity found is due to carbonate, bicarbonate, or hydroxide) can be made using calculations based on the values given in Table 8.1.

**TABLE 8.1**
**Interpretation of Results Values**

| | Alkalinity (mg/L as $CaCO_3$) | | |
|---|---|---|---|
| **Results of Titration** | **Bicarbonate Alkalinity** | **Carbonate Alkalinity** | **Hydroxide Alkalinity** |
| P = 0 | T | 0 | 0 |
| P = T | 0 | 0 | T |
| P < 1/2T | T – 2P | 2P | 0 |
| P = 1/2T | 0 | T | 0 |
| P > 1/2T | 0 | 2(T – P) | 2P –T |

*Source:* APHA, *Standard Methods*, Vol. 19, American Public Health Association, Washington, DC, 1995, p. 2-28.

*Note:* P, phenolphthalein alkalinity; T, total alkalinity.

## ■ Example 8.11

*Problem:* A water sample is tested for phenolphthalein and total alkalinity. If the phenolphthalein alkalinity is 10 mg/L as $CaCO_3$ and the total alkalinity is 52 mg/L as $CaCO_3$, determine the bicarbonate, carbonate, and hydroxide alkalinity of the water.

*Solution:* Based on titration test results, P alkalinity (10 mg/L) is less than half of the T alkalinity (52 mg/L ÷ 2 = 26 mg/L; see Table 8.1). Therefore, each type of alkalinity is calculated as follows:

$$\text{Bicarbonate alkalinity} = \text{T} - 2\text{P} = 52 \text{ mg/L} - 2(10 \text{ mg/L}) = 32 \text{ mg/L as CaCO}_3$$

$$\text{Carbonate alkalinity} = 2\text{P} = 2(10 \text{ mg/L}) = 20 \text{ mg/L as CaCO}_3$$

$$\text{Hydroxide alkalinity} = 0 \text{ mg/L as CaCO}_3$$

## ■ Example 8.12

*Problem*: Determine the phenolphthalein, total bicarbonate, carbonate, and hydroxide alkalinity based on the following alkalinity titrations on a water sample:

Sample, 100 mL
1.4 mL titrant used to pH 8.3
2.4 mL total titrant used to pH 4.5.
Acid normality, 0.02-$N$ $H_2SO_4$

*Solution:*

$$\text{Phenolphthalein alkalinity} = \frac{A \times N \times 50{,}000}{\text{Sample (mL)}} = \frac{1.4 \text{ mL} \times 0.02\ N \times 50{,}000}{100 \text{ mL}} = 14 \text{ mg/L as CaCO}_3$$

$$\text{Total alkalinity} = \frac{B \times N \times 50{,}000}{\text{Sample (mL)}} = \frac{2.4 \text{ mL} \times 0.02\ N \times 50{,}000}{100 \text{ mL}} = 24 \text{ mg/L as CaCO}_3$$

Now use Table 8.1 to calculate the other alkalinity constituents (P > 1/2T):

$$\text{Bicarbonate alkalinity} = 0 \text{ mg/L as CaCO}_3$$

$$\text{Carbonate alkalinity} = 2\text{T} - 2\text{P} = 2(24 \text{ mg/L}) - 2(14 \text{ mg/L}) = 20 \text{ mg/L as CaCO}_3$$

$$\text{Hydroxide alkalinity} = 2\text{P} - \text{T} = 2(14 \text{ mg/L}) - (24 \text{ mg/L}) = 4 \text{ mg/L as CaCO}_3$$

## LIME DOSAGE CALCULATION FOR REMOVAL OF CARBONATE HARDNESS

The lime–soda ash water-softening process uses lime ($Ca(OH)_2$) and soda ash ($Na_2CO_3$) to precipitate hardness from solution. Carbonate hardness (calcium and magnesium bicarbonates) is complexed by lime. Noncarbonate hardness (calcium and magnesium sulfates or chlorides) requires the addition of soda ash for precipitation. The molecular weights of various compounds used in lime–soda softening calculations are as follows:

Quicklime ($CaO$) = 56
Hydrated lime ($Ca(OH)_2$) = 74
Magnesium ($Mg^{2+}$) = 24.3
Carbon dioxide ($CO_2$) = 44
Magnesium hydroxide ($Mg(OH)_2$) = 58.3
Soda ash ($Na_2CO_3$) = 100
Alkalinity (as $CaCO_3$) = 100
Hardness ($CaCO_3$) = 100

To calculate the quicklime or hydrated lime dosage (in mg/L), use Equation 8.10:

$$\text{Quicklime (CaO) dosage (mg/L)} = \frac{(A+B+C+D)\times 1.15}{\%\ \text{Purity of lime}/100} \tag{8.10}$$

where

$A$ = $CO_2$ in source water (mg/L as $CO_2$) × (56/44).
$B$ = Bicarbonate alkalinity removed in softening (mg/L as $CaCO_3$) × (56/100).
$C$ = Hydroxide alkalinity in softener effluent (mg/L as $CaCO_3$) × (56/100).
$D$ = Magnesium removed in softening (mg/L as $Mg^{2+}$) × (56/24.3).
1.15 = Excess lime dosage (15% excess).

***Note:*** For hydrated lime dosage, use Equation 8.10 but substitute 74 for 56 in *A*, *B*, *C*, and *D*.

### ■ Example 8.13

*Problem:* A water sample has a carbon dioxide content of 4 mg/L as $CO_2$, total alkalinity of 130 mg/L as $CaCO_3$, and magnesium content of 26 mg/L as $Mg^{2+}$. Approximately how much quicklime (CaO) (90% purity) will be required for softening? (Assume 15% excess lime.)

*Solution:* Calculate the *A* to *D* factors:

$A$ = (mg/L as $CO_2$) × (56/44) = 4 mg/L × (56/44) = 5.1 mg/L
$B$ = (mg/L as $CaCO_3$) × (56/100) = 130 mg/L × (56/100) = 73 mg/L
$C$ = 0 mg/L
$D$ = (mg/L as $Mg^{2+}$) × (56/24.3) = 26 mg/L × (56/24.3) = 60 mg/L

Calculate the estimated quicklime dosage:

$$\text{Quicklime dosage} = \frac{(A + B + C + D) \times 1.15}{\%\ \text{Purity of lime}/100}$$

$$= \frac{(5\ \text{mg/L} + 73\ \text{mg/L} + 0 + 60\ \text{mg/L}) \times 1.15}{0.90}$$

$$= 176\ \text{mg/L CaO}$$

### ■ Example 8.14

*Problem:* The characteristics of a water sample are as follows: 4 mg/L $CO_2$ as $CO_2$, 175 mg/L total alkalinity as $CaCO_3$, and 20 mg/L magnesium as $Mg^{2+}$. What is the estimated hydrated lime (Ca $(OH)_2$) (90% pure) dosage in mg/L required for softening? (Assume 15% excess lime.)

*Solution:* Determine the $A$ to $D$ factors:

$A$ = (mg/L as $CO_2$) × (74/44) = 4 mg/L × (74/44) = 7 mg/L
$B$ = (mg/L as $CaCO_3$) × (74/100) = 175 mg/L × (74/100) = 130 mg/L
$C$ = 0 mg/L
$D$ = (mg/L as $Mg^{2+}$) × (74/24.3) = 20 mg/L × (74/24.3) = 61 mg/L

Calculate the estimated hydrated lime dosage:

$$\text{Hydrated lime dosage} = \frac{(A + B + C + D) \times 1.15}{\%\ \text{Purity of lime}/100}$$

$$= \frac{(7\ \text{mg/L} + 130\ \text{mg/L} + 0 + 61\ \text{mg/L}) \times 1.15}{0.90}$$

$$= 253\ \text{mg/L Ca(OH)}_2$$

## CALCULATION FOR REMOVAL OF NONCARBONATE HARDNESS

Soda ash is used for precipitation and removal of noncarbonate hardness. To calculate the soda ash dosage required, we use Equations 8.11 and 8.12.

$$\text{Total hardness (mg/L as CaCO}_3) = \text{Carbonate hardness (mg/L as CaCO}_3) + \text{Noncarbonate hardness (mg/L as CaCO}_3) \quad (8.11)$$

$$\text{Soda ash (mg/L)} = \text{Noncarbonate hardness (mg/L as CaCO}_3) \times (106/100) \quad (8.12)$$

### ■ Example 8.15

*Problem:* A water sample has a total hardness of 250 mg/L as $CaCO_3$ and a total alkalinity of 180 mg/L. What soda ash dosage (mg/L) will be required to remove the noncarbonate hardness?

*Solution:* Calculate the noncarbonate hardness:

$$\text{Total hardness (mg/L as CaCO}_3) = \text{Carbonate hardness (mg/L as CaCO}_3) + \text{Noncarbonate hardness (mg/L as CaCO}_3)$$

$$250 \text{ mg/L} = 180 \text{ mg/L} + x \text{ mg/L}$$

$$250 \text{ mg/L} - 180 \text{ mg/L} = x \text{ mg/L}$$

$$70 \text{ mg/L} = x$$

Calculate the soda ash required:

$$\text{Soda ash} = \text{Noncarbonate hardness (mg/L as CaCO}_3) \times (106/100)$$

$$= 70 \text{ mg/L} \times (106/100)$$

$$= 74.2 \text{ mg/L}$$

### ■ Example 8.16

*Problem:* Calculate the soda ash required (in mg/L) to soften water if the water has a total hardness of 192 mg/L and a total alkalinity of 103 mg/L.

*Solution*: Determine noncarbonate hardness:

$$\text{Total hardness (mg/L as CaCO}_3) = \text{Carbonate hardness (mg/L as CaCO}_3) + \text{Noncarbonate hardness (mg/L as CaCO}_3)$$

$$192 \text{ mg/L} = 103 \text{ mg/L} + x \text{ mg/L}$$

$$192 \text{ mg/L} - 103 \text{ mg/L} = x \text{ mg/L}$$

$$89 \text{ mg/L} = x$$

Calculate soda ash required:

$$\text{Soda ash} = \text{Noncarbonate hardness (mg/L as CaCO}_3) \times (106/100)$$

$$= 89 \text{ mg/L} \times (106/100)$$

$$= 94 \text{ mg/L}$$

## RECARBONATION CALCULATION

Recarbonation involves the reintroduction of carbon dioxide into the water, either during or after lime softening, lowering the pH of the water to about 10.4. After the addition of soda ash, recarbonation lowers the pH of the water to about 9.8, promoting better precipitation of calcium carbonate and magnesium hydroxide. Equations 8.13 and 8.14 are used to estimate carbon dioxide dosage:

$$\text{Excess lime (mg/L)} = (A + B + C + D) \times 0.15 \tag{8.13}$$

$$\text{Total } CO_2 \text{ dosage (mg/L)} = \left[Ca(OH)_2 \text{ excess (mg/L)} \times 44/74\right] + \left[Mg^{2+} \text{ residual (mg/L)} \times 44/24.3\right] \tag{8.14}$$

### ■ Example 8.17

*Problem:* The *A*, *B*, *C*, and *D* factors of the excess lime equation have been calculated as follows: $A = 14$ mg/L, $B = 126$ mg/L, $C = 0$, $D = 66$ mg/L. If the residual magnesium is 5 mg/L, how much carbon dioxide (in mg/L) is required for recarbonation?

*Solution:* Calculate the excess lime concentration:

$$\text{Excess lime} = (A + B + C + D) \times 0.15$$
$$= (14 \text{ mg/L} + 126 \text{ mg/L} + 0 + 66 \text{ mg/L}) \times 0.15 = 31 \text{ mg/L}$$

Determine the required carbon dioxide dosage:

$$\text{Total } CO_2 \text{ dosage} = [31 \text{ mg/L} \times (44/74)] + [(5 \text{ mg/L} \times (44/24.3)]$$
$$= 18 \text{ mg/L} + 9 \text{ mg/L} = 27 \text{ mg/L}$$

### ■ Example 8.18

*Problem:* The *A*, *B*, *C*, and *D* factors of the excess lime equation have been calculated as follows: $A = 10$ mg/L, $B = 87$ mg/L, $C = 0$, $D = 111$ mg/L. If the residual magnesium is 5 mg/L, what carbon dioxide dosage would be required for recarbonation?

*Solution:* Calculate the excess lime concentration:

$$\text{Excess lime} = (A + B + C + D) \times 0.15$$
$$= (10 \text{ mg/L} + 87 \text{ mg/L} + 0 + 111 \text{ mg/L}) \times 0.15 = 31 \text{ mg/L}$$

Determine the required carbon dioxide dosage:

$$\text{Total } CO_2 \text{ dosage} = [31 \text{ mg/L} \times (44/74)] + [(5 \text{ mg/L} \times (44/24.3)]$$
$$= 18 \text{ mg/L} + 9 \text{ mg/L} = 27 \text{ mg/L}$$

## CALCULATING FEED RATES

The appropriate chemical dosage for various unit processes is typically determined by lab or pilot-scale testing (e.g., jar testing, pilot plant), monitoring, and historical experience. When the chemical dosage is determined, the feed rate can be calculated by Equation 8.15. When the chemical feed rate is known, this value must be translated into a chemical feeder setting.

$$\text{Feed rate (lb/day)} = \text{Flow rate (MGD)} \times \text{Dose (mg/L)} \times 8.34 \text{ lb/gal} \tag{8.15}$$

To calculate the lb/min chemical required we use Equation 8.16.

$$\text{Chemical (lb/min)} = \frac{\text{Chemical (lb/day)}}{1440 \text{ min/day}} \tag{8.16}$$

### ■ Example 8.19

*Problem:* Jar tests indicate that the optimum lime dosage is 200 mg/L. If the flow to be treated is 4.0 MGD, what should be the chemical feeder settings in lb/day and lb/min?

*Solution*: Calculate the lb/day feed rate using Equation 8.15:

$$\text{Feed rate} = \text{Flow rate (MGD)} \times \text{Dose (mg/L)} \times 8.34 \text{ lb/gal}$$

$$= 4.0 \text{ MGD} \times 200 \text{ mg/L} \times 8.34 \text{ lb/gal} = 6672 \text{ lb/day}$$

Convert this feed rate to lb/min:

$$\text{Chemical (lb/min)} = \frac{\text{Chemical (lb/day)}}{1440 \text{ min/day}} = \frac{6672 \text{ lb/day}}{1440 \text{ min/day}} = 4.6 \text{ lb/min}$$

### ■ Example 8.20

*Problem:* What should the lime dosage settings be in lb/day and lb/hr if the optimum lime dosage has been determined to be 125 mg/L and the flow to be treated is 1.1 MGD?

*Solution:* The lb/day feed rate for lime is

$$\text{Feed rate} = \text{Flow rate (MGD)} \times \text{Dose (mg/L)} \times 8.34 \text{ lb/gal}$$

$$= 1.1 \text{ MGD} \times 125 \text{ mg/L} \times 8.34 \text{ lb/gal} = 1147 \text{ lb/day}$$

Convert this to a lb/hr feed rate:

$$\text{Chemical (lb/hr)} = \frac{\text{Chemical (lb/day)}}{24 \text{ hr/day}} = \frac{1147 \text{ lb/day}}{24 \text{ hr/day}} = 48 \text{ lb/hr}$$

## ION EXCHANGE CAPACITY

An ion exchange softener is a common alternative to the use of lime and soda ash for softening water. Natural water sources contain dissolved minerals that dissociate in water to form charged particles called ions. Of main concern are the positively charged ions of calcium, magnesium, and sodium; bicarbonate, sulfate, and chloride are the normal negatively charged ions of concern. An ion exchange medium, called *resin*, is a material that will exchange a hardness-causing ion for another one that does not cause hardness, hold the new ion temporarily, and then release it when a regenerating solution is poured over the resin. The removal capacity of an exchange resin is generally reported as grains of hardness removal per cubic foot of resin (gr/ft$^3$ resin). To calculate the removal capacity of the softener, we use Equation 8.17.

$$\text{Exchange capacity (gr)} = \text{Removal capacity (gr/ft}^3\text{)} \times \text{Media volume (ft}^3\text{)} \quad (8.17)$$

### ■ EXAMPLE 8.21

*Problem:* The hardness removal capacity of an exchange resin is 24,000 gr/ft$^3$. If the softener contains a total of 70 ft$^3$ of resin, what is the total exchange capacity (grains) of the softener?

*Solution:*

$$\text{Exchange capacity} = \text{Removal capacity (gr/ft}^3\text{)} \times \text{Media volume (ft}^3\text{)}$$
$$= 22{,}000 \text{ gr/ft}^3 \times 70 \text{ ft}^3 = 1{,}540{,}000 \text{ gr}$$

### ■ EXAMPLE 8.22

*Problem:* An ion exchange water softener has a diameter of 7 ft. The depth of the resin is 5 ft. If the resin has a removal capacity of 22 kilograins (Kgr)/ft$^3$, what is the total exchange capacity of the softener (in grains)?

*Solution:* Before the exchange capacity of a softener can be calculated, the ft$^3$ resin volume must be known:

$$\text{Volume (ft}^3\text{)} = 0.785 \times D^2 \times \text{Depth (ft)} = 0.785 \times (7 \text{ ft})^2 \times 5 \text{ ft} = 192 \text{ ft}^3$$

Now calculate the exchange capacity of the softener:

$$\text{Exchange capacity} = \text{Removal capacity (gr/ft}^3\text{)} \times \text{Media volume (ft}^3\text{)}$$
$$= 22{,}000 \text{ gr/ft}^3 \times 192 \text{ ft}^3 = 4{,}224{,}000 \text{ gr}$$

## WATER TREATMENT CAPACITY

To calculate when the resin must be regenerated (based on volume of water treated), we must know the exchange capacity of the softener and the hardness of the water. Equation 8.18 is used for this calculation.

$$\text{Water treatment capacity (gal)} = \frac{\text{Exchange capacity (gr)}}{\text{Hardness (gr/gal)}} \tag{8.18}$$

### ■ Example 8.23

*Problem:* An ion exchange softener has an exchange capacity of 2,445,000 gr. If the hardness of the water to be treated is 18.6 gr/gal, how many gallons of water can be treated before regeneration of the resin is required?

*Solution:*

$$\text{Water treatment capacity} = \frac{\text{Exchange capacity (gr)}}{\text{Hardness (gr/gal)}} = \frac{2{,}455{,}000 \text{ gr}}{18.6 \text{ gr/gal}} = 131{,}989 \text{ gal}$$

### ■ Example 8.24

*Problem:* An ion exchange softener has an exchange capacity of 5,500,000 gr. If the hardness of the water to be treated is 14.8 gr/gal, how many gallons of water can be treated before regeneration of the resin is required?

*Solution:*

$$\text{Water treatment capacity} = \frac{\text{Exchange capacity (gr)}}{\text{Hardness (gr/gal)}} = \frac{5{,}500{,}000 \text{ gr}}{14.8 \text{ gr/gal}} = 371{,}622 \text{ gal}$$

### ■ Example 8.25

*Problem:* The hardness removal capacity of an ion exchange resin is 25 Kgr/ft³. The softener contains a total of 160 ft³ of resin. If the water to be treated contains 14.0 gr/gal hardness, how many gallons of water can be treated before regeneration of the resin is required?

*Solution:* Both the water hardness and the exchange capacity of the softener must be determined before the gallons of water can be calculated.

$$\begin{aligned}\text{Exchange capacity} &= \text{Removal capacity (gr/ft}^3) \times \text{Media volume (ft}^3) \\ &= 25{,}000 \text{ gr/ft}^3 \times 160 \text{ ft}^3 = 4{,}000{,}000 \text{ gr}\end{aligned}$$

Calculate the gallons water treated:

$$\text{Water treatment capacity} = \frac{\text{Exchange capacity (gr)}}{\text{Hardness (gr/gal)}} = \frac{4{,}000{,}000 \text{ gr}}{14.0 \text{ gr/gal}} = 285{,}714 \text{ gal}$$

## TREATMENT TIME CALCULATIONS (UNTIL REGENERATION REQUIRED)

After calculating the total number of gallons of water to be treated (before regeneration), we can also calculate the operating time required to treat that amount of water. Equation 8.19 is used to make this calculation.

$$\text{Operating time (hr)} = \frac{\text{Water treated (gal)}}{\text{Flow rate (gph)}} \tag{8.19}$$

### ■ Example 8.26

*Problem:* An ion exchange softener can treat a total of 642,000 gal before regeneration is required. If the flow rate treated is 25,000 gph, how many hours of operation remain before regeneration is required?

*Solution:*

$$\text{Operating time} = \frac{\text{Water treated (gal)}}{\text{Flow rate (gph)}} = \frac{642{,}000 \text{ gal}}{25{,}000 \text{ gph}} = 25.7 \text{ hr}$$

### ■ Example 8.27

*Problem:* An ion exchange softener can treat a total of 820,000 gal of water before regeneration of the resin is required. If the water is to be treated at a rate of 32,000 gph, how many hours of operation remain until regeneration is required?

*Solution:*

$$\text{Operating time} = \frac{\text{Water treated (gal)}}{\text{Flow rate (gph)}} = \frac{820{,}000 \text{ gal}}{32{,}000 \text{ gph}} = 25.6 \text{ hr}$$

## SALT AND BRINE REQUIRED FOR REGENERATION

When calcium and magnesium ions replace the sodium ions in the ion exchange resin, the resin can no longer remove the hardness ions from the water. When this occurs, pumping a concentrated solution (10 to 14% sodium chloride solution) on the resin will regenerate the resin. When the resin is completely recharged with sodium ions, it is ready for softening again. Typically, the salt dosage required to prepare the brine solution ranges from 5 to 15 lb salt per $ft^3$ resin. Equation 8.20 is used to calculate the salt required (lb), and Equation 8.21 is used to calculate the brine (gal).

$$\text{Salt required (lb)} = \text{Salt required (lb/Kgr removed)} \times \text{Hardness removed (Kgr)} \tag{8.20}$$

**TABLE 8.2**
**Salt Solutions Table**

| % NaCl | NaCl (lb/gal) | NaCl (lb/ft³) |
|---|---|---|
| 10 | 0.87 | 6.69 |
| 11 | 0.99 | 7.41 |
| 12 | 1.09 | 8.14 |
| 13 | 1.19 | 8.83 |
| 14 | 1.29 | 9.63 |
| 15 | 1.39 | 10.4 |

$$\text{Brine (gal)} = \frac{\text{Salt required (lb)}}{\text{Brine solution (lb salt per gal brine)}} \tag{8.21}$$

To determine the brine solution (Equation 8.21), we must refer to a salt solutions table (Table 8.2).

## ■ Example 8.28

*Problem*: An ion exchange softener removes 1,310,000 gr hardness from the water before the resin must be regenerated. If 0.3 lb salt is required for each kilograin removed, how many pounds of salt will be required to prepare the brine to be used in resin regeneration?

*Solution:*

$$\text{Salt required} = \text{Salt required (lb/Kgr removed)} \times \text{Hardness removed (Kgr)}$$

$$= 0.3 \text{ lb/Kgr removed} \times 1310 \text{ Kgr} = 393 \text{ lb salt}$$

## ■ Example 8.29

*Problem:* A total of 430 lb salt will be required to regenerate an ion exchange softener. If the brine solution is to be 12% (see Table 8.2 to determine the lb salt per gal brine for a 12% brine solution), how many gallons brine will be required?

*Solution:*

$$\text{Brine (12\%)} = \frac{\text{Salt required (lb)}}{\text{Brine solution (lb salt per gal brine)}}$$

$$= \frac{430 \text{ lb}}{1.09 \text{ lb salt per gal brine}} = 394 \text{ gal}$$

Thus, it takes 430 lb salt to make up the 394 gal brine required for the desired 12% brine solution.

# 9 Water Treatment Practice Calculations

1. The static water level for a well is 91 ft. If the pumping water level is 98 ft, what is the well drawdown?

2. The static water level for a well is 110 ft. The pumping water level is 125 ft. What is the well drawdown?

3. Before the pump is started, the water level is measured at 144 ft. The pump is then started. If the pumping water level is determined to be 161 ft, what is the well drawdown?

4. The static water level of a well is 86 ft. The pumping water level is determined using a sounding line. The air pressure applied to the sounding line is 3.7 psi and the length of the sounding line is 112 ft. What is the drawdown?

5. A sounding line is used to determine the static water level for a well. The air pressure applied is 4.6 psi and the length of the sounding line is 150 ft. If the pumping water level is 171 ft, what is the drawdown?

6. If the well yield is 300 gpm and the drawdown is measured to be 20 ft, what is the specific capacity?

7. During a 5-minute well yield test, 420 gal were pumped from the well. What is the well yield in gpm?

8. Once the drawdown of a well stabilized, it was determined that the well produced 810 gal during a 5-minute pumping test. What is the well yield in gpm?

9. During a test for well yield, 856 gal were pumped from the well. If the well yield test lasted 5 minutes, what was the well yield in gpm? In gph?

10. A bailer is used to determine the approximate yield of a well. The bailer is 12 ft long and has a diameter of 12 in. If the bailer is placed in the well and removed 12 times during a 5-minute test, what is the well yield in gpm?

11. During a 5-minute well yield test, 750 gal of water were pumped from the well. At this yield, if the pump is operated 10 hr each day, how many gallons of water are pumped daily?

12. The discharge capacity of a well is 200 gpm. If the drawdown is 28 ft, what is the specific yield in gpm/ft of drawdown?

13. A well produces 620 gpm. If the drawdown for the well is 21 ft, what is the specific yield in gpm/ft of drawdown?

14. A well yields 1100 gpm. If the drawdown is 41.3 ft, what is the specific yield in gpm/ft of drawdown?

15. The specific yield of a well is listed as 33.4 gpm/ft. If the drawdown for the well is 42.8 ft, what is the well yield in gpm?

16. A new well is to be disinfected with chlorine at a dosage of 40 mg/L. If the well casing diameter is 6 in. and the length of the water-filled casing is 140 ft, how may pounds of chlorine will be required?

17. A new well with a casing diameter of 12 in. is to be disinfected. The desired chlorine dosage is 40 mg/L. If the casing is 190 ft long and the water level in the well is 81 ft from the top of the well, how many pounds of chlorine will be required?

18. An existing well has a total casing length of 210 ft. The top 180 ft of the casing has a 12-in. diameter and the bottom 40 ft of the casing has an 8-in. diameter. The water level is 71 ft from the top of the well. How many pounds of chlorine will be required if a chlorine dosage of 110 mg/L is desired?

19. The water-filled casing of a well has a volume of 540 gal. If 0.48 lb of chlorine were used in disinfection, what was the chlorine dosage (in mg/L)?

20. For the disinfection of a well, 0.09 lb of chlorine is required. If sodium hypochlorite (5.25% available chlorine) is to be used, how many fluid ounces of sodium hypochlorite are required?

21. A new well is to be disinfected with calcium hypochlorite (65% available chlorine). The well casing diameter is 6 in. and the length of the water-filled casing is 120 ft. If the desired chlorine dosage is 50 g/L, how many ounces (dry measure) of calcium hypochlorite will be required?

22. How many pounds of chloride of lime (25% available chlorine) will be required to disinfect a well if the casing is 18 inches in diameter and 200 ft long with a water level at 95 ft from the top of the well? The desired chlorine dosage is 100 mg/L.

23. The water-filled casing of a well has a volume of 240 gal. How many fluid ounces of sodium hypochlorite (5.25% available chlorine) are required to disinfect the well if a chlorine concentration of 60 mg/L is desired?

24. The pressure gauge reading at a pump discharge head is 4.0 psi. What is this discharge head expressed in feet?

25. The static water level of a well is 94 ft. The well drawdown is 24 ft. If the gauge reading at the pump discharge head is 3.6 psi, what is the field head?

26. The pumping water level for a well is 180 ft. The discharge pressure measured at the pump discharge head is 4.2 psi. If the pump capacity is 800 gpm, what is the water horsepower?

27. The pumping water level for a well is 200 ft. The pump discharge head is 4.4 psi. If the pump capacity is 1000 gpm, what is the water horsepower?

28. The bowl head of a vertical turbine pump is 184 ft and the bowl efficiency is 83%. If the capacity of the vertical turbine pump is 700 gpm, what is the bowl horsepower?

29. A vertical turbine pump has a bowl horsepower of 59.5 bhp. The shaft is 1-1/4 inches in diameter and rotates at a speed of 1450 rpm. If the shaft is 181 ft long, what is the field bhp? Shaft friction loss is 0.67.

30. The field brake horsepower for a deep-well turbine pump is 58.3 bhp. The thrust bearing loss is 0.5 hp. If the motor efficiency provided by the manufacturer is 90%, what is the horsepower input to the motor?

31. The total brake horsepower for a deep-well turbine pump is 56.4 bhp. If the water horsepower is 45 whp, what is the field efficiency?

32. The total brake horsepower for a pump is 55.7 bhp. If the motor is 90% efficient and the water horsepower is 43.5 whp, what is the overall efficiency of the unit?

33. A pond has an average length of 400 ft, average width of 110 ft, and estimated average depth of 14 ft. What is the estimated volume of the pond in gallons?

34. A pond has an average length of 400 ft and an average width of 110 ft. If the maximum depth of the pond is 30 ft, what is the estimated gallon volume of the pond?

35. A pond has an average length of 200 ft, an average width of 80 ft, and an average depth of 12 ft. What is the acre-foot volume of the pond?

36. A small pond has an average length of 320 ft, an average width of 170 ft, and a maximum depth of 16 ft. What is the acre-foot volume of the pond?

37. For algae control in a reservoir, a dosage of 0.5-mg/L copper is desired. The reservoir has a volume of 20 MG. How many pounds of copper sulfate pentahydrate (25% available copper) will be required?

38. The desired copper dosage in a reservoir is 0.5 mg/L. The reservoir has a volume of 62 ac-ft. How many pounds of copper sulfate pentahydrate (25% available copper) will be required?

39. A pond has a volume of 38 ac-ft. If the desired copper sulfate dosage is 1.1 lb copper sulfate per ac-ft, how many pounds of copper sulfate will be required?

40. A pond has an average length of 250 ft, an average width of 75 ft, and an average depth of 10 ft. If the desired dosage is 0.8 lb copper sulfate per ac-ft, how many pounds of copper sulfate will be required?

41. A storage reservoir has an average length of 500 and an average width of 100 ft. If the desired copper sulfate dosage is 5.1 lb copper sulfate per acre, how many pounds of copper sulfate will be required?

42. The static water level for a well is 93.9 ft. If the pumping water level is 131.5 ft, what is the drawdown?

43. During a 5-minute well yield test, 707 gal were pumped from the well. What is the well yield in gpm? In gph?

44. A bailer is used to determine the approximate yield of a well. The bailer is 12 ft long and has a diameter of 12 in. If the bailer is placed in the well and removed 8 times during a 5-minute test, what is the well yield in gpm?

45. The static water level in a well is 141 ft. The pumping water level is determined using the sounding line. The air pressure applied to the sounding line is 3.5 psi and the length of the sounding line is 167 ft. What is the drawdown?

46. A well produces 610 gpm. If the drawdown for the well is 28 ft, what is the specific yield in gpm/ft of drawdown?

47. A new well is to be disinfected with a chlorine dose of 55 mg/L. If the well casing diameter is 6 in. and the length of the water-filled casing is 150 ft, how many pounds of chlorine will be required?

48. During a 5-minute well yield test, 780 gal of water were pumped from the well. At this yield, if the pump is operated 8 hr each day, how many gallons of water are pumped daily?

49. The water-filled casing of a well has a volume of 610 gal. If 0.47 lb of chlorine was used for disinfection, what was the chlorine dosage (in mg/L)?

50. An existing well has a total casing length of 230 ft. The top 170 ft of the casing has a 12-in. diameter, and the bottom 45 ft of the casing has an 8-in. diameter. The water level is 81 ft from the top of the well. How many pounds of chlorine will be required if a chlorine dosage of 100 mg/L is desired?

51. The disinfection of a well requires 0.3 lb of chlorine. If sodium hypochlorite is to be used (5.25% available chlorine), how many fluid ounces of sodium hypochlorite are required?

52. The pressure gauge reading at a pump discharge head is 4.5 psi. What is this discharge head expressed in feet?

53. The static water level of a well is 95 ft. The well drawdown is 25 ft. If the gauge reading at the pump discharge head is 3.6 psi, what is the field head?

54. The pumping water level for a well is 191 ft. The discharge pressure measured at the pump discharge head is 4.1 psi. If the pump capacity is 850 gpm, what is the horsepower?

55. A deep-well vertical turbine pump delivers 800 gpm. The bowl head is 175 ft and the bowl efficiency is 80%. What is the bowl horsepower?

56. The field brake horsepower for a deep-well turbine pump is 47.8 bhp. The thrust bearing loss is 0.8 hp. If the motor efficiency provided by the manufacturer is 90%, what is the horsepower input to the motor?

57. The total brake horsepower for a deep-well turbine is 57.4 bhp. If the water horsepower is 45.6 whp, what is the field efficiency?

58. The total brake horsepower for a pump is 54.7 bhp. If the motor is 90% efficient and the water horsepower is 44.6 whp, what is the overall efficiency of the unit?

59. The desired copper dosage at a reservoir is 0.5 mg/L. The reservoir has a volume of 53 ac-ft. How many pounds of copper sulfate pentahydrate (25% available copper) will be required?

60. A storage reservoir has an average length of 440 ft and an average width of 140 ft. If the desired copper sulfate dosage is 5.5 lb copper sulfate per acre, how many pounds of copper sulfate will be required?

61. A flash mix chamber is 4 ft wide by 5 ft long, with water to a depth of 3 ft. What is the gallon volume of water in the flash mix chamber?

62. A flocculation basin is 50 ft long by 20 ft wide, with water to a depth of 8 ft. What is the gallon volume of water in the basin?

63. A flocculation basin is 40 ft long by 16 ft wide, with water to a depth of 8 ft. How many gallons of water are in the basin?

64. A flash mix chamber is 5 ft square, with water to a depth of 42 in. What is the gallon volume of water in the flash mixing chamber?

65. A flocculation basin is 25 ft wide by 40 ft long and contains water to a depth of 9 ft, 2 in. What is the gallon volume of water in the flocculation basin?

66. The flow to a flocculation basin is 3,625,000 gpd. If the basin is 60 ft long by 25 ft wide with water to a depth of 9 ft, what is the detention time (in minutes) of the flocculation basin?

67. A flocculation basin is 50 ft long by 20 ft wide and has a water level of 8 ft. What is the detention time (in minutes) in the basin if the flow to the basin is 2.8 MGD?

68. A flash mix chamber 6 ft long by 5 ft wide and 5 ft deep receives a flow of 9 MGD. What is the detention time (in seconds) in the chamber?

69. A flocculation basin is 50 ft long by 20 ft wide and has a water depth of 10 ft. If the flow to the basin is 2,250,000 gpd, what is the detention time in minutes?

70. A flash mix chamber is 4 ft square, with a water depth of 42 in. If the flash mix chamber receives a flow of 3.25 MGD, what is the detention time in seconds?

71. The desired dry alum dosage, as determined by the jar test, is 10 mg/L. Determine the lb/day setting on a dry alum feeder if the flow is 3,450,000 gpd.

72. Jar tests indicate that the best polymer dose for a water sample is 12 mg/L. If the flow to be treated is 1,660,000 gpd, what should be the dry chemical feed setting (in lb/day)?

73. Determine the desired lb/day setting on a dry alum feeder if jar tests indicate an optimum dose of 10 mg/L and the flow to be treated is 2.66 MGD.

74. The desired dry alum dose is 9 mg/L, as determined by a jar test. If the flow to be treated is 940,000 gpd, How many pounds per day dry alum will be required?

75. A flow of 4.10 MGD is to be treated with a dry polymer. If the desired dose is 12 mg/L, what should be the dry chemical feeder setting (lb/day)?

76. Jar tests indicate that the best alum dose for a unit process is 7 mg/L. The flow to be treated is 1.66 MGD. Determine the gpd setting for the alum solution feeder if the liquid alum contains 5.24 lb of alum per gallon of solution.

77. The flow to a plant is 3.43 MGD. Jar testing indicates that the optimum alum dose is 12 mg/L. What should the gpd setting be for the solution feeder if the alum solution is a 55% solution?

78. Jar tests indicate that the best alum dose for a unit process is 10 mg/L. The flow to be treated is 4.13 MGD. Determine the gpd setting for the alum solution feeder if the liquid alum contains 5.40 lb of alum per gallon of solution.

79. Jar tests indicate that the best liquid alum dose for a unit process is 11 mg/L. The flow to be treated is 880,000 gpd. Determine the gpd setting for the liquid alum chemical feeder if the liquid alum is a 55% solution.

80. A flow of 1,850,000 gpd is to be treated with alum. Jar tests indicate that the optimum alum dose is 10 mg/L. If the liquid alum contains 640 mg alum per mL solution, what should be the gpd setting for the alum solution feeder?

81. The desired solution feed rate was calculated to be 40 gpd. What is this feed rate expressed as mL/min?

82. The desired solution feed rate was calculated to be 34.2 gpd. What is this feed rate expressed as mL/min?

83. The optimum polymer dose has been determined to be 10 mg/L. The flow to be treated is 2,880,000 gpd. If the solution to be used contains 55% active polymer, what should be the solution chemical feeder setting (in mL/min)?

84. The optimum polymer dose for a 2,820,000-gpd flow has been determined to be 6 mg/L. If the polymer solution contains 55% active polymer, what should be the solution chemical feeder setting (in mL/min)? Assume that the polymer solution weighs 8.34 lb/gal.

85. Jar tests indicate that the best alum dose for a unit process is 10 mg/L. The liquid alum contains 5.40 alum per gallon of solution. What should the setting be on the solution chemical feeder (in mL/min) when the flow to be treated is 3.45 MGD?

86. If 140 g of dry polymer are dissolved in 16 gal of water, what is the percent strength of the solution? (1 g = 0.0022 lb.)

87. If 22 oz. of dry polymer are added to 24 gal of water, what is the percent strength (by weight) of the polymer solution?

88. How many gallons of water must be added to 2.1 lb dry alum to make a 0.8% solution?

89. An 11% liquid polymer is to be used in making up a polymer solution. How many pounds of liquid polymer should be mixed with water to produce 160 lb of a 0.5% polymer solution?

90. An 8% polymer solution is used in making up a solution. How many gallons of liquid polymer should be added to the water to make up 50 gal of a 0.2% polymer solution? The liquid polymer has a specific gravity of 1.3. Assume that the polymer solution has a specific gravity of 1.0.

91. How many gallons of an 11% liquid polymer should be mixed with water to produce 80 gal of a 0.8% polymer solution? The density of the polymer liquid is 10.1 lb/gal. Assume that the density of the polymer solution is 8.34 lb/gal.

92. If 32 lb of a 10% strength solution are mixed with 66 lb of a 0.5% strength solution, what is the percent strength of the solution mixture?

93. If 5 gal of a 15% strength solution are added to 40 gal of a 0.20% strength solution, what is the percent strength of the solution mixture? Assume that the 15% strength solution weighs 11.2 lb/gal and the 0.20% solution weighs 8.34 lb/gal.

94. If 12 gal of a 12% strength solution are mixed with 50 gal of a 0.75% strength solution, what is the percent strength of the solution mixture? Assume that the 12% solution weighs 10.5 lb/gal and the 0.75% solution weighs 8.40 lb/gal.

95. Calculate the actual chemical feed rate (in lb/day) if a container is placed under a chemical feeder and 2.3 lb are collected during a 30-minute period.

96. Calculate the actual chemical feed rate (in lb/day), if a container is placed under a chemical feeder and 42 oz. are collected during a 45-minute period.

97. To calibrate a chemical feeder, a container is first weighed (14 oz.) and then placed under the chemical feeder. After 30 minutes the container is weighed again. If the weight of the container with chemical is 2.4 lb, what is the actual chemical feed rate (in lb/day)?

98. A chemical feeder is to be calibrated. The container to be used to collect the chemical is placed under the chemical feeder and weighed (0.6 lb). After 30 minutes, the weight of the container was found to be 2.8 lb. Based on this test, what is the actual chemical feed rate (in lb/day)?

99. During a 24-hr period, a flow of 1,920,000 gpd water is treated. If 42 lb of polymer were used for coagulation during that 24-hr period, what is the polymer dosage (in mg/L)?

100. A calibration test is conducted for a solution chemical feeder. During a 24-hour period the solution feeder delivers 70 gal of solution. The polymer solution is a 1.6% solution. What is the lb/day solution feed rate? Assume that the polymer solution weighs 8.34 lb/gal.

101. A calibration test is conducted for a solution chemical feeder. During a 5-minute test, the pump delivered 590 mL of a 1.2% polymer solution. The specific gravity of the polymer solution is 1.09. What is the polymer dosage rate in lb/day?

102. During a 5-minute calibration test for a solution chemical feeder, the solution feeder delivers 725 mL. The polymer solution is a 1.2% solution. What is the lb/day polymer feed rate? Assume that the polymer solution weighs 8.34 lb/gal.

103. A solution chemical feeder delivered 950 mL of solution during a 5-minute calibration test. The polymer solution is a 1.4% strength solution. What is the polymer dosage rate in lb/day? Assume that the polymer solution weighs 8.34 lb/gal.

104. If 1730 mL of a 1.9% polymer solution are delivered during a 10-minute calibration test, and the polymer solution has a specific gravity of 1.09, what is the polymer feed rate in lb/day?

105. A pumping rate calibration test is conducted for a 5-minute period. The liquid level in the 4-ft-diameter solution tank is measured before and after the test. If the level drops 4 in. during the 5-minute test, what is the gpm pumping rate?

106. During a 15-minute pumping rate calibration test, the liquid level in the 4-ft-diameter solution tank drops 4 in. What is the pumping rate in gpm?

107. The liquid level in a 3-ft-diameter solution tank drops 3 in. during a 10-minute pumping rate calibration test. What is the gpm pumping rate? Assuming a continuous pumping rate, what is the gpd pumping rate?

108. During a 15-minute pumping rate calibration test, the solution level in the 3-ft-diameter chemical tank dropped 2 in. Assume that the pump operates at that rate during the next 24 hr. If the polymer solution is a 1.3% solution, what is the lb/day polymer feed? Assume that the polymer solution weighs 8.34 lb/gal.

109. The level in a 4 ft-diameter chemical tank drops 2 in. during a 30-minute pumping rate calibration test. The polymer solution is 1.45% solution. If the pump operates at the same rate for 24 hr, what is the polymer feed in lb/day?

110. The amount of chemical used for each day during a week is given below. Based on this information, what was the average lb/day chemical use during the week?

     Monday, 81 lb/day
     Tuesday, 73 lb/day
     Wednesday, 74 lb/day
     Thursday, 66 lb/day
     Friday, 79 lb/day
     Saturday, 80 lb/day
     Sunday, 82 lb/day

111. The average chemical use at a plant is 90 lb/day. If the chemical inventory in stock is 2200 lb, how many days supply is this?

112. The chemical inventory in stock is 889 lb. If the average chemical use at a plant is 58 lb/day, how many days supply is this?

113. The average gallons of polymer solution used each day at a treatment plant is 88 gpd. A chemical feed tank has a diameter of 3 ft and contains solution to a depth of 3 ft, 4 in. How many days supply are represented by the solution in the tank?

114. Jar tests indicate that the optimum polymer dose for a unit process is 2.8 mg/L. If the flow to be treated is 1.8 MGD, how many pounds of dry polymer will be required for a 30-day period?

115. A flash mix chamber 4 ft long by 4 ft wide with a 3-ft water depth receives a flow of 6.1 MGD. What is the detention time in the chamber (in seconds)?

116. A flocculation basin is 50 ft long by 20 ft wide and has a water depth of 9 ft. What is the volume of water in the basin (in gallons)?

117. The desired dry alum dosage, as determined by jar testing, is 9 mg/L. Determine the lb/day setting on a dry alum feeder if the flow is 4.35 MGD.

118. The flow to a plant is 3.15 MGD. Jar testing indicates that the best alum dose is 10 mg/L. What should the gpd setting be for the solution feeder if the alum solution is a 50% solution? Assume that the alum solution weighs 8.34 lb/gal.

119. A flash mix chamber is 4 ft square with a water depth of 2 ft. What is the gallon volume of water in this chamber?

120. The desired solution feed rate was calculated to be 45 gpd. What is this feed rate expressed as mL/min?

121. A flocculation basin is 40 ft long by 20 ft wide and has a water depth of 9 ft, 2 in. If the flow to the basin is 2,220,000 gpd, what is the detention time in minutes?

122. The optimum polymer dose has been determined to be 8 mg/L. The flow to be treated is 1,840,000 gpd. If the solution to be used contains 60% active polymer, what should be the solution chemical feeder setting in mL/min? The polymer solution weighs 10.2 lb/gal.

123. The desired solution feed rate was calculated to be 180 gpd. What is this feed rate expressed as mL/min?

124. Determine the desired lb/day setting on a dry alum feeder if jar tests indicate an optimum dose of 6 mg/L and the flow to be treated is 925,000 gpd.

125 How many gallons of water must be added to 2.7 lb of dry alum to make a 1.4% solution?

126. If 25 lb of a 16% strength solution are mixed with 140 lb of a 0.6% strength solution, what is the percent strength of the solution mixture?

127. Calculate the chemical feed rate in lb/day if a container is placed under a chemical feeder and 4.0 lb of chemical are collected during a 30-min period.

128. A chemical feeder is to be calibrated. The container to be used to collect chemical is placed under the chemical feeder and weighed (2.0 lb). After 30 minutes, the weight of the container and chemical is found to be 4.2 lb. Based on this test, what is the chemical feed rate in lb/day?

129. If 190 g of dry polymer are dissolved in 25 gal of water, what is the percent strength (by weight) of the solution? (1 g = 0.0022 lb.)

130. During a 5-minute calibration test for a solution chemical feeder, the feeder delivers 760 mL. The polymer solution is a 2.0% solution. What is the lb/day polymer feed rate? Assume that the polymer solution weighs 8.34 lb/gal.

131. Jar tests indicate that the best alum dose for a unit process is 14 mg/L. The flow to be treated is 4.2 MG. Determine the gpd setting for the alum solution feeder if the liquid alum contains 5.66 lb of alum per gallon of solution.

132. A 10% liquid polymer is to be used in making up a polymer solution. How many pounds of liquid polymer should be mixed with water to produce 210 lb of a 0.8% polymer solution?

133. How many pounds of a 60% polymer solution and water should be mixed together to form 175 lb of a 1% solution?

134. During a 10-minute pumping rate calibration test, the liquid level in the 4-ft-diameter solution tank drops 3 in. What is the pumping rate in gpm?

135. How many gallons of a 10% liquid polymer should be mixed with water to produce 80 gal of a 0.6% polymer solution? The density of the polymer liquid is 10.2 lb/gal. Assume that the density of the polymer solution is 8.34 lb/gal.

136. A calibration test is conducted of a solution chemical feeder. During a 5-minute test, the pump delivered 710 mL of a 0.9% polymer solution. The specific gravity of the polymer solution is 1.3. What is the polymer dosage rate in lb/day?

137. Jar tests indicate that the best polymer dose for a unit process is 6 mg/L. If the flow to be treated is 3.7 MGD, at this rate how many pounds of dry polymer will be required for a 30-day period?

138. The chemical inventory in stock is 550 lb. If the average chemical use at the plant is 80 lb/day, how many days supply is this?

139. A sedimentation basin is 70 ft long by 30 ft wide. If the water depth is 14 ft, what is the volume of water in the tank in gallons?

140. A circular clarifier has a diameter of 80 ft. If the water depth is 12 ft, how many gallons of water are in the tank?

141. A sedimentation tank is 70 ft long by 20 ft wide and has water to a depth of 10 ft. What is the volume of water in the tank in gallons?

142. A sedimentation basin is 40 ft long by 25 ft wide. When the basin contains 50,000 gallons, what would be the water depth?

143. A circular clarifier is 75 ft in diameter. If the water depth is 10 ft, 5 in., what is the volume of water in the clarifier in gallons?

144. A rectangular sedimentation basin is 70 ft long by 25 ft wide and has water to a depth of 10 ft. The flow to the basin is 2,220,000 gpd. Calculate the detention time in hours for the sedimentation basin.

145. A circular clarifier has a diameter of 80 ft and an average water depth of 12 ft. If the flow to the clarifier is 2,920,000 gpd, what is the detention time in hours?

146. A rectangular sedimentation basin is 60 ft long by 20 ft wide with water to a depth of 10 ft. If the flow to the basin is 1,520,000 gpd, what is the detention time in hours?

147. A circular clarifier has a diameter of 60 ft and an average water depth of 12 ft. What flow rate (in MGD) corresponds to a detention time of 3 hr?

148. A sedimentation basin is 70 ft long by 25 ft wide. The average water depth is 12 ft. If the flow to the basin is 1,740,000 gpd, what is the detention time in the sedimentation basin in hours?

149. A rectangular sedimentation basin is 60 ft long by 25 ft wide. When the flow is 510 gpm, what is the surface overflow rate in gpm/ft$^2$?

150. A circular clarifier has a diameter of 70 ft. If the flow to the clarifier is 1610 gpm, what is the surface overflow rate in gpm/ft$^2$?

151. A rectangular sedimentation basin receives a flow of 540,000 gpd. If the basin is 50 ft long by 20 ft wide, what is the surface overflow rate in gpm/ft$^2$?

152. A sedimentation basin is 80 ft long by 25 ft wide. To maintain a surface overflow rate of 0.5 gpm/ft$^2$, what is the maximum flow to the basin in gpd?

153. A circular clarifier 60 ft in diameter receives a flow of 1,820,000 gpd. What is the surface overflow rate in gpm/ft$^2$?

154. A sedimentation basin is 80 ft long by 25 ft wide and operates at a depth of 12 ft. If the flow to the basin is 1,550,000 gpd, what is the mean flow velocity in fpm?

155. A sedimentation basin is 70 ft long by 30 ft wide and operates at a depth of 12 ft. If the flow to the basin is 1.8 MGD, what is the mean flow velocity in fpm?

156. A sedimentation basin is 80 ft long by 30 ft wide. The water level is 14 ft. When the flow to the basin is 2.45 MGD, what is the mean flow velocity in fpm?

157. The flow to a sedimentation basin is 2,880,000 gpd. If the length of the basin is 70 ft, the width of the basin is 40 ft, and the depth of water in the basin is 10 ft, what is the mean flow velocity in fpm?

158. A sedimentation basin 50 ft long by 25 ft wide receives a flow of 910,000 gpd. The basin operates at a depth of 10 ft. What is the mean flow velocity in the basin in ft/min?

159. A circular clarifier receives a flow of 2,520,000 gpd. If the diameter of the weir is 70 ft, what is the weir loading rate in gpm/ft?

160. A rectangular sedimentation basin has 170 ft of weir. If the flow to the basin is 1,890,000 gpd, what is the weir loading rate in gpm/ft?

161. A rectangular sedimentation basin has 120 ft of weir. If the flow over the weir is 1,334,000 gpd, what is the weir loading rate in gpm/ft?

162. A circular clarifier receives a flow of 3.7 MGD. If the diameter of the weir is 70 ft, what is the weir loading rate in gpm/ft?

163. A rectangular sedimentation basin has 160 ft of weir. If the flow over the weir is 1.9 MGD, what is the weir loading rate in gpm/ft?

164. A 100-mL sample of slurry from a solids contact unit is placed in a graduated cylinder and allowed to settle for 10 minutes. The settled sludge at the bottom of the graduated cylinder after 10 minutes is 22 mL. What is the percent settled sludge of the sample?

165. A 100-mL sample of slurry from a solids contact unit is placed in a graduated cylinder. After 10 minutes, 25 mL of sludge settled to the bottom of the cylinder. What is the percent settled sludge of the sample?

166. A percent settled sludge test is conducted on a 100-mL sample of solids contact unit slurry. After 10 minutes of settling, 15-mL of sludge settled to the bottom of the cylinder. What is the percent settled sludge of the sample?

167. A 100-mL sample of slurry from a solids contact unit is placed in a graduated cylinder. After 10 minutes, 16 mL of sludge settled to the bottom of the cylinder. What is the percent settled sludge of the sample?

168. Raw water requires an alum dose of 52 mg/L, as determined by jar testing. If a residual 40 mg/L alkalinity must be present in the water to promote precipitation of the alum added, what is the total alkalinity required in mg/L? (1 mg/L alum reacts with 0.45 mg/L alkalinity.)

169. Jar tests indicate that a level of 60 mg/L alum is optimum for a particular raw water unit. If a residual 30 mg/L alkalinity must be present to promote complete precipitation of the alum added, what is the total alkalinity required in mg/L? (1 mg/L alum reacts with 0.45 mg/L alkalinity.)

170. Suppose 40 mg/L alkalinity is required to react with alum and ensure complete precipitation. If the raw water has an alkalinity of 26 mg/L as bicarbonate, how may mg/L alkalinity should be added to the water?

171. If 40 mg/L alkalinity is required to react with alum and ensure complete precipitation of the alum added, and the raw water has an alkalinity of 28 mg/L as bicarbonate, how many mg/L alkalinity must be added to the water?

172. If 15 mg/L alkalinity must be added to raw water, how many mg/L lime will be required to provide this amount of alkalinity? (1 mg/L alum reacts with 0.45 mg/L alkalinity and 1 mg/L alum reacts with 0.45 mg/L lime.)

173. It has been calculated that 20 mg/L alkalinity must be added to a raw water unit. How many mg/L lime will be required to provide this amount of alkalinity? (1 mg/L alum reacts with 0.45 mg/L alkalinity and 1 mg/L alum reacts with 0.35 mg/L lime.)

174. Given the following data, calculate the required lime dose in mg/L:

    Alum dose required per jar tests = 55 mg/L
    1 mg/L alum reacts with 0.45 mg/L alkalinity
    Raw water alkalinity = 35 mg/L
    1 mg/L alkalinity reacts with 0.35 mg/L lime
    Residual alkalinity required for precipitation= 30 mg/L

175. The lime dose for a raw water unit has been calculated to be 13.8 mg/L. If the flow to be treated is 2.7 MGD, how many lb/day of dry lime will be required?

176. The lime dose for a solids contact unit has been calculated to be 12.3 mg/L. If the flow to be treated is 2,2400,000 gpd, how many lb/day lime will be required?

177. The flow to a solids contact clarifier is 990,000 gpd. If the lime dose required is determined to be 16.1 mg/L, how many lb/day lime will be required?

178. A solids contact clarification unit receives a flow of 2.2 MGD. Alum is to be used for coagulation purposes. If a lime dose of 15 mg/L will be required, how many lb/day lime is this?

179. If 205 lb/day lime are required to raise the alkalinity of the water passing through a solids contact clarifier, how many g/min lime does this represent? (1 lb = 453.6 g.)

180. A lime dose of 110 lb/day is required for a raw water unit feeding a solids contact clarifier. How many g/min lime does this represent? (1 lb = 453.6 g.)

181. A lime dose of 12 mg/L is required to raise the pH of a particular water. If the flow to be treated is 900,000 gpd, what g/min lime dose will be required? (1 lb = 453.6 g.)

182. A lime dose of 14 mg/L is required to raise the alkalinity of a unit process. If the flow to be treated is 2,660,000 gpd, what is the g/min lime dose required? (1 lb = 452.6 g.)

183. A rectangular sedimentation basin is 65 ft long by 30 ft wide with water to a depth of 12 ft. If the flow to the basin is 1,550,000 gpd, what is the detention time in hours?

184. A sedimentation basin is 70 ft long by 30 ft wide. If the water depth is 14 ft, what is the volume of water in the tank in gallons?

185. A sedimentation basin 65 ft long by 25 ft wide operates at a depth of 12 ft. If the flow to the basin is 1,620,000 gpd, what is the mean flow velocity in fpm?

186. A rectangular sedimentation basin receives a flow of 635,000 gpd. If the basin is 40 ft long by 25 ft wide, what is the surface overflow rate in $gpm/ft^2$?

187. A circular clarifier has a diameter of 70 ft. If the water depth is 14 ft, how many gallons of water are in the tank?

188. A rectangular sedimentation basin has 180 ft of weir. If the flow to the basin is 2,220,000 gpd, what is the weir loading rate in gpm/ft?

189. A circular clarifier has a diameter of 60 ft and an average water depth of 10 ft. If the flow to the clarifier is 2.56 MGD, what is the detention time in hours?

190. A sedimentation basin 55 ft long by 30 ft wide operates at a depth of 12 ft. If the flow to the basin is 1.75 MGD, what is the mean flow velocity in fpm?

191. A circular clarifier has a diameter of 70 ft. If the flow to the clarifier is 1700 gpm, what is the surface overflow rate in $gpm/ft^2$?

192. A circular clarifier receives a flow of 3.15 MGD. If the diameter of the weir is 70 ft, what is the weir loading rate in gpm/ft?

193. A circular clarifier has a diameter of 60 ft and an average water depth of 12 ft. What flow rate (MGD) corresponds to a detention time of 2 hr?

194. The flow to a sedimentation basin is 3.25 MGD. If the length of the basin is 80 ft, the width of the basin is 30 ft, and the depth of water in the basin is 14 ft, what is the mean flow velocity in ft/min?

195. A 100-mL sample of slurry from a solids contact clarification unit is placed in a graduated cylinder and allowed to settle for 10 minutes. The settled sludge at the bottom of the graduated cylinder after 10 minutes is 26 mL. What is the percent settled sludge of the sample?

196. A raw water unit requires an alum dose of 50 mg/L, as determined by jar testing. If a residual 30 mg/L alkalinity must be present in the water to promote complete precipitation of the alum added, what is the total alkalinity required in mg/L? (1 mg/L alum reacts with 0.45 mg/L alkalinity.)

197. A sedimentation basin is 80 ft long by 30 ft wide. To maintain a surface overflow rate of 0.7 $gpm/ft^2$, what is the maximum flow to the basin in gpd?

198. The lime dose for a raw water unit has been calculated to be 14.5 mg/L. If the flow to be treated is 2,410,000 gpd, how many lb/day lime will be required?

199. A 100-mL sample of slurry from a solids contact clarification unit is placed in a graduated cylinder. After 10 minutes, 21 mL of sludge had settled to the bottom of the cylinder. What is the percent settled sludge of the sample?

200. A circular clarifier receives a flow of 3.24 MGD. If the diameter of the weir is 80 ft, what is the weir loading rate in gpm/ft?

201. Suppose 50 mg/L alkalinity are required to react with alum and ensure complete precipitation of the alum added. If the raw water has an alkalinity of 30 mg/L as bicarbonate, how many mg/L alkalinity should be added to the water?

202. Given the following data, calculate the required lime dose in mg/L:

   Alum dose required per jar tests = 50 mg/L
   1 mg/L alum reacts with 0.45 mg/L alkalinity
   Raw water alkalinity = 33 mg/L
   1 mg/L alkalinity reacts with 0.35 mg/L lime
   Residual alkalinity required for precipitation = 30 mg/L

203. To raise the alkalinity of the water passing through a solids contact clarifier, 192 lb/day lime will be required. How many g/min lime does this represent? (1 lb = 453.6 g.)

204. A solids contact clarification unit receives a flow of 1.5 MGD. Alum is to be used for coagulation purposes. If a lime dose of 16 mg/L is required, how many lb/day lime is this?

205. A lime dose of 14 mg/L is required to raise the alkalinity of a raw water unit. If the flow to be treated is 2,880,000 gpd, what is the g/min lime dose required?

206. During an 80-hr filter run, 14.2 million gal of water are filtered. What is the average gpm flow rate through the filter during this time?

207. The flow rate through a filter is 2.97 MGD. What is this flow rate expressed as gpm?

208. At an average flow rate through a filter of 3200 gpm, how long a filter run (in hours) would be required to produce 16 MG of filtered water?

209. The influent valve to a filter is closed for a 5-minute period. During this time, the water level in the filter drops 12 in. If the filter is 45 ft long by 22 ft wide, what is the gpm flow rate through the filter?

210. A filter is 40 ft long by 30 ft wide. To verify the flow rate through the filter, the filter influent valve is closed for a 5-minute period and the water drop is measured. If the water level in the filter drops 14 in. during the 5 minutes, what is the gpm flow rate through the filter?

211. The influent valve to a filter is closed for 6 minutes. The water level in the filter drops 18 in. during the 6 minutes. If the filter is 35 ft long by 18 ft wide, what is the gpm flow rate through the filter?

212. A filter 20 ft long by 18 ft wide receives a flow of 1760 gpm. What is the filtration rate in $gpm/ft^2$?

213. A filter has a surface area of 32 ft by 18 ft. If the filter receives a flow a 2,150,000 gpd, what is the filtration rate in $gpm/ft^2$?

214. A filter 38 ft long by 24 ft wide produces 18.1 MG during a 71.6-hr filter run. What is the average filtration rate for this filter run?

215. A filter 33 ft long by 24 ft wide produces 14.2 MG during a 71.4-hr filter run. What is the average filtration rate for this filter run?

216. A filter 38 ft long by 22 ft wide receives a flow of 3.550,000 gpd. What is the filtration rate in $gpm/ft^2$?

217. A filter is 38 ft long by 18 ft wide. During a test of filter flow rate, the influent valve to the filter is closed for 5 minutes. The water level drops 22 in. during this period. What is the filtration rate for the filter in $gpm/ft^2$?

218. A filter is 33 ft long by 24 ft wide. During a test of flow rate, the influent valve to the filter is closed for 6 minutes. The water level drops 21 in. during this period. What is the filtration rate for the filter in $gpm/ft^2$?

219. The total water filtered between backwashes is 2.87 MG. If the filter is 20 ft by 18 ft, what is the unit filter run volume in $gal/ft^2$?

220. The total amount of water filtered during a filter run was 4,180,000 gal. If the filter is 32 ft long by 20 ft wide, what is the UFRV in gal/$ft^2$?

221. During a particular filter run, 2,980,000 gal of water were filtered. If the filter is 24 ft long by 18 ft wide, what is the UFRV in gal/$ft^2$?

222. The average filtration rate for a filter is 3.4 gpm/$ft^2$. If the filter run time was 3330 minutes, what was the unit filter run volume in gal/$ft^2$?

223. A filter ran 60.5 hr between backwashes. If the average filtration rate during that time was 2.6 gpm/$ft^2$, what was the UFRV in gal/$ft^2$?

224. A filter with a surface area of 380 $ft^2$ has a backwash flow rate of 3510 gpm. What is the filter backwash rate in gpm/$ft^2$?

225. A filter 18 ft long by 14 ft wide has a backwash flow rate of 3580 gpm. What is the filter backwash rate in gpm/$ft^2$?

226. A filter has a backwash rate of 16 gpm/$ft^2$. What is this backwash rate expressed as in./min?

227. A filter 30 ft by 18 ft has a backwash flow rate of 3650 gpm. What is the filter backwash rate in gpm/$ft^2$?

228. A filter 18 ft long by 14 ft wide has a backwash rate of 3080 gpm. What is this backwash rate expressed as in./min rise?

229. A backwash flow rate of 6650 gpm for a total backwashing period of 6 minutes would require how many gallons of water for backwashing?

230. For a backwash flow rate of 9100 gpm and a total backwash time of 7 minutes, how many gallons of water will be required for backwashing?

231. How many gallons of water would be required to provide a backwash flow rate of 4670 gpm for 5 min?

232. A backwash flow rate of 6750 gpm for 6 minutes would require how many gallons of water?

233. If 59,200 gal of water will be required to provide a 7-minute backwash of a filter, what depth of water is required in the backwash water tank to provide this backwashing capability? The tank has a diameter of 40 ft.

234. The volume of water required for backwashing has been calculated to be 62,200 gal. What is the required depth of water in the backwash water tank to provide this amount of water if the diameter of the tank is 52 ft?

235. If 42,300 gal of water will be required for backwashing a filter, what depth of water is required in the backwash water tank to provide this much water? The diameter of the tank is 42 ft.

236. A backwash rate of 7150 gpm is desired for a total backwash time of 7 minutes. What depth of water is required in the backwash water tank to provide this much water? The diameter of the tank is 40 ft.

237. A backwash rate of 8860 gpm is desired for a total backwash time of 6 minutes. What depth of water is required in the backwash water tank to provide this backwashing capability? The diameter of the tank is 40 ft.

238. A filter is 42 ft long by 22 ft wide. If the desired backwash rate is 19 gpm/ft$^2$, what backwash pumping rate (gpm) will be required?

239. The desired backwash pumping rate for a filter is 20 gpm/ft$^2$. If the filter is 36 ft long by 26 ft wide, what backwash pumping rate (gpm) will be required?

240. A filter is 22 ft square. If the desired backwash rate is 16 gpm/ft$^2$, what backwash pumping rate (in gpm) will be required?

241. The desired backwash pumping rate for a filter is 24 gpm/ft$^2$. If the filter is 26 ft long by 22 ft wide, what backwash pumping rate (gpm) will be required?

242. If 17,100,000 gal of water are filtered during a filter run and 74,200 gal of this product water are used for backwashing, what percent of the product water is used for backwashing?

243. Suppose 6.10 MG of water were filtered during a filter run. If 37,200 gal of this product water were used for backwashing, what percent of the product water was used for backwashing?

244. If 59,400 gal of product water are used for filter backwashing at the end of a filter run and 13,100,000 gal are filtered during the filter run, what percent of the product water is used for backwashing?

245. If 11,110,000 gal of water are filtered during a particular filter run and 52,350 gal of product water are used for backwashing, what percent of the product water is used for backwashing?

246. A 3625-mL sample of filter media was taken for mud ball evaluation. The volume of water in the graduated cylinder rose from 600 mL to 635 mL when mud balls were placed in the cylinder. What is the percent mud ball volume of the sample?

247. Five samples of filter media are taken for mud ball evaluation. The volume of water in the graduated cylinder rose from 510 mL to 535 mL when mud balls were placed in the cylinder. What is the percent mud ball volume of the sample? The mud ball sampler has a volume of 705 mL.

248. A filter media is tested for the presence of mud balls. The mud ball sampler has a total sample volume of 705 mL. Five samples were taken from the filter. When the mud balls were placed in 520 mL of water, the water level rose to 595 mL. What is the percent mud ball volume of the sample?

249. Five samples of media filter are taken and evaluated for the presence of mud balls. The volume of water in the graduated cylinder rises from 520 mL to 562 mL when the mud balls are placed in the water. What is the percent mud ball volume of the sample? The mud ball sampler has a volume of 705 mL.

250. During an 80-hr filter run, 11.4 million gal of water are filtered. What is the average gpm flow rate through the filter during this time?

251. A filter is 40 ft long by 25 ft wide. If the filter receives a flow of 3.56 MGD, what is the filtration rate in gpm/ft$^2$?

252. The total water filtered between backwashes is 2.88 MG. If the filter is 25 ft by 25 ft, what is the unit filter run volume in gal/ft$^2$?

253. At an average flow rate through a filter of 2900 gpm, how long a filter run (in hours) would be required to produce 14.8 MG of filtered water?

254. A filter is 38 ft long by 26 ft wide. To verify the flow rate through the filter, the filter influent valve is closed for a period of 5 minutes and the water drop is measured. If the water level in the filter drops 14 in. during the 5-minute period, what is the gpm flow rate through the filter?

255. If 3,450,000 gal of water are filtered during a particular filter run, and the filter is 30 ft long by 25 ft wide, what is the unit filter run volume in gal/ft$^2$?

256. A filter 30 ft long by 20 ft wide produces 13,500,000 gal during a 73.8-hr filter run. What is the average filtration rate (in gpm/ft$^2$) for this filter run?

257. A filter with a surface area of 360 ft$^2$ has a backwash flow rate of 3220 gpm. What is the filter backwash rate in gpm/ft$^2$?

258. A backwash flow rate of 6350 gpm for a total backwashing period of 6 minutes would require how many gallons of water for backwashing?

259. The influent valve to a filter is closed for a 5-minute period. During this time, the water level in the filter drops 14 in. If the filter is 30 ft long by 22 ft wide, what is the filtration rate in gpm/ft$^2$?

260. If 53,200 gal of water are required to provide a 6-minute backwash of a filter, what depth of water is required in the backwash water tank to provide this backwashing capability? The tank has a diameter of 45 ft.

261. The average filtration rate for a filter is 3.3 gpm/ft$^2$. If the filter run time was 3620 minutes, what was the unit filter run volume in gal/ft$^2$?

262. A filter is 40 ft long by 25 ft wide. During a test of filter flow rate, the influent valve to the filter is closed for 5 minutes. The water level drops 20 in. during this period. What is the filtration rate for the filter in gpm/ft$^2$?

263. A filter 35 ft by 25 ft has a backwash flow rate of 3800 gpm. What is the filter backwash rate in gpm/ft$^2$?

264. How many gallons of water would be required to provide a backwash flow rate of 4500 gpm for 7 minutes?

265. The desired backwash pumping rate for a filter is 16 gpm/ft$^2$. If the filter is 30 ft long by 30 ft wide, what backwash pumping rate (in gpm) will be required?

266. A filter 25 ft long by 20 ft wide has a backwash rate of 2800 gpm. What is this backwash rate expressed as in./min rise?

267. A filter is 30 ft long by 25 ft wide. During a test of flow rate, the influent valve to the filter is closed for 6 minutes. The water level drops 18 in. during this period. What is the filtration rate for the filter in gpm/$ft^2$?

268. A filter is 45 ft long by 25 ft wide. If the desired backwash rate is 18 gpm/$ft^2$, what backwash pumping rate (gpm) will be required?

269. If 18,200,000 gal of water are filtered during a filter run and 71,350 gal of this product water are used for backwashing, what percent of the product water is used for backwashing?

270. If 86,400 gal of water are required for backwashing a filter, what depth of water is required in the backwash water tank to provide this much water? The diameter of the tank is 35 ft.

271. A 3480-mL sample of filter media was taken for mud ball evaluation. The volume of water in the graduated cylinder rose from 500 mL to 527 mL when the mud balls were placed in the cylinder. What is the percent mud ball volume of the sample?

272. If 51,200 gal of product water are used for filter backwashing at the end of a filter run and 13.8 MG are filtered during the filter run, what percent of the product water is used for backwashing?

273. Five samples of filter media are taken for mud ball evaluation. The volume of water in the graduated cylinder rose from 500 mL to 571 mL when the mud balls were placed in the cylinder. What is the percent mud ball volume of the sample? The mud ball sampler has a volume of 695 mL.

274. A 20 ft by 25 ft sand filter is set to operate for 36 hr before being backwashed. Backwashing takes 15 minutes and uses 12 gpm/$ft^2$ until the water is clear. The water in the backwash storage tank drops 25 ft during the backwash. Each filter run produces 3.7 MG. (a) What is the filtration rate in gpm/$ft^2$? (b) How many gallons of backwash water were used? (c) What was the percentage of product water used for backwashing? (d) If the initial tank water depth was 70 ft, how deep is the water after backwashing? (e) What is the UFRV?

275. Determine the chlorinator setting (lb/day) required to treat a flow of 3.5 MGD with a chlorine dose of 1.8 mg/L.

276. A flow of 1,340,000 gpd is to receive a chlorine dose of 2.5 mL. What should the chlorinator settling be in lb/day?

277. A pipeline 12 inches in diameter and 1200 ft long is to be treated with a chlorine dose of 52 mg/L. How many pounds of chlorine will this require?

278. A chlorinator setting is 43 lb per 24 hr. If the flow being treated is 3.35 MGD, what is the chlorine dosage expressed as mg/L?

279. The flow totalizer reading at 9 a.m. on Thursday was 18,815,108 gal and at 9 a.m. on Thursday was 19,222,420 gal. If the chlorinator setting was 16 lb for this 24-hr period, what is the chlorine dosage in mg/L?

280. The chlorine demand of a water process is 1.6 mg/L. If the desired chlorine residual is 0.5 mg/L, what is the desired chlorine dose in mg/L?

281. The chlorine dosage for a water process is 2.9 mL. If the chlorine residual after a 30-minute contact time is found to be 0.7 mg/L, what is the chlorine demand expressed in mg/L?

282. A flow of 3,850,000 gpd is to be disinfected with chlorine. If the chlorine demand is 2.6 mg/L and a chlorine residual of 0.8 mg/L is desired, what should be the chlorinator setting in lb/day?

283. A chlorinator setting is increased by 6 lb/day. The chlorine residual before the increased dosage was 0.4 mg/L. After the increased dose, the chlorine residual was 0.8 mg/L. The average flow rate being treated is 1,100,000 gpd. Is the water being chlorinated beyond the breakpoint?

284. A chlorinator setting of 19 lb of chlorine per 24 hr results in a chlorine residual of 0.4 mg/L. The chlorinator setting is increased to 24 lb per 24 hr. The chlorine residual increases to 0.5 mg/L at this new dosage rate. The average flow being treated is 2,100,000 gpd. On the basis of this information, is the water being chlorinated past the breakpoint?

285. A chlorine dose of 48 lb/day is required to treat a particular water unit. If calcium hypochlorite (65% available chlorine) is to be used, how many lb/day hypochlorite will be required?

286. A chlorine dose of 42 lb/day is required to disinfect a flow of 2,220,000 gpd. If the calcium hypochlorite to be used contains 65% available chlorine, how many lb/day hypochlorite will be required?

287. A water flow of 928,000 gpd requires a chlorine dose of 2.7 mg/L. If calcium hypochlorite (65% available chlorine) is to be used, how many lb/day hypochlorite are required?

288. If 54 lb of hypochlorite (65% available chlorine) are used in a day and the flow rate treated is 1,512,000 gpd, what is the chlorine dosage in mg/L?

289. A flow of 3,210,000 gpd is disinfected with calcium hypochlorite (65% available chlorine). If 49 lb of hypochlorite are used in a 24-hr period, what is the mg/L chlorine dosage?

290. If 36 lb/day sodium hypochlorite are required for disinfection of a 1.7-MGD flow, how many gallons per day sodium hypochlorite is this?

291. Hypochlorite is used to disinfect the water pumped from a well. The hypochlorite solution contains 3% available chlorine. A chlorine dose of 2.2 mg/L is required for adequate disinfection throughout the distribution system. If the flow from the well is 245,000 gpd, how much sodium hypochlorite (in gpd) will be required?

292. The water from a well is disinfected by a hypochlorinator. The flow totalizer indicates that 2,330,000 gal of water were pumped during a 70-day period. The 3% sodium hypochlorite solution that was used to treat the well water is pumped from a 3-ft-diameter storage tank. During the 7-day period, the level in the tank dropped 2 ft, 10 in. What is the chlorine dosage in mg/L?

293. A hypochlorite solution (4% available chlorine) is used to disinfect a water unit. A chlorine dose of 1.8 mg/L is desired to maintain an adequate chlorine residual. If the flow being treated is 400 gpm, what hypochlorite solution flow (in gpd) will be required?

294. A sodium hypochlorite solution (3% available chlorine) is used to disinfect the water pumped from a well. A chlorine dose of 2.9 mg/L is required to provide adequate disinfection. How many gallons per day of sodium hypochlorite will be required if the flow being chlorinated is 955,000 gpd?

295. If 22 lb of calcium hypochlorite (65% available chlorine) are added to 60 gal of water, what is the percent chlorine (by weight) of the solution?

296. If 320 g of calcium hypochlorite are dissolved in 7 gal of water, what is the percent chlorine (by weight) of the solution? (1 g = 0.0022 lb.)

297. How many pounds of dry hypochlorite (65% available chlorine) must be added to 65 gal of water to make a 3% chlorine solution?

298. A 10% liquid hypochlorite is to be used in making up a 2% hypochlorite solution. How many gallons of liquid hypochlorite should be mixed with water to produce 35 gal of a 2% hypochlorite solution?

299. How many gallons of 13% liquid hypochlorite should be mixed with water to produce 110 gal of a 1.2% hypochlorite solution?

300. If 6 gal of a 12% sodium hypochlorite solution are added to a 55-gal drum, how much water should be added to produce a 2% hypochlorite solution?

301. If 50 lb of a hypochlorite solution (11% available chlorine) are mixed with 220 lb of another hypochlorite solution (1% available chlorine), what is the percent chlorine of the solution mixture?

302. If 12 gal of a 12% hypochlorite solution are mixed with 60 gal of a 1.5% hypochlorite solution, what is the percent strength of the solution mixture?

303. If 16 gal of a 12% hypochlorite solution are added to 70 gal of a 1% hypochlorite solution, what is the percent strength of the solution mixture?

304. The average calcium hypochlorite use at a plant is 44 lb/day. If the chemical inventory in stock is 1000 lb, how many days supply is this?

305. The average daily use of sodium hypochlorite solution at a plant is 80 gpd. A chemical feed tank has a diameter of 4 ft and contains solution to a depth of 3 ft, 8 inches. How many days supply is represented by the solution in the tank?

306. An average of 24 lb of chlorine is used each day at a plant. How many pounds of chlorine would be used in one week if the hour meter on the pump registered 150 hr of operation that week?

307. A chlorine cylinder has 91 lb of chlorine at the beginning of a week. The chlorinator setting is 12 lb per 24 hr. If the pump hour meter indicates the pump has operated 111 hours during the week, how many lb chlorine should be in the cylinder at the end of the week?

308. An average of 55 lb of chlorine is used each day at a plant. How many 150-lb chlorine cylinders will be required each month? Assume a 30-day month.

309. The average sodium hypochlorite use at a plant is 52 gpd. If the chemical feed tank is 3 ft in diameter, how many feet should the solution level in the tank drop in 2 days?

310. The chlorine demand of a water unit is 1.8 mg/L. If the desired chlorine residual is 0.9 mg/L, what is the desired chlorine dose in mg/L?

311. Determine the chlorinator setting (lb/day) required to treat a flow of 980,000 gpd with a chlorine dose of 2.3 mg/L.

312. A chlorine dose of 60 lb/day is required to treat a water unit. If calcium hypochlorite (65% available chlorine) is to be used, how many pounds/day of hypochlorite will be required?

313. If 51 lb/day sodium hypochlorite are required for disinfection of a 2.28-MGD flow, how many gallons per day sodium hypochlorite is this?

314. The chlorine dosage for a water unit is 3.1 mg/L. If the chlorine residual after a 30-minute contact time is found to be 0.6 mg/L, what is the chlorine demand expressed in mg/L?

315. When 30 lb of calcium hypochlorite (65% available chlorine) are added to 66 gal of water, what is the percent chlorine (by weight) of the solution?

316. What chlorinator setting is required to treat a flow of 1620 gpm with a chlorine dose of 2.8 mg/L?

317. A chlorine dose of 2.8 mg/L is required for adequate disinfection of a water unit. If a flow of 1.33 MGD is treated, how many gpd of sodium hypochlorite are required? The sodium hypochlorite contains 12.5% available chlorine.

318. A pipeline 8 inches in diameter and 1600 ft long is to be treated with a chlorine dose of 60 mg/L. How many pounds of chlorine will this require?

319. A chlorinator setting of 15 lb of chlorine per 24 hr results in a chlorine residual of 0.5 mg/L. The chlorinator setting is increased to 18 lb per 24 hr. The chlorine residual increases to 0.6 mg/L at this new dosage rate. The average flow being treated is 2,110,000 gpd. On the basis of this information, is the water being chlorinated past the breakpoint?

320. If 70 gal of a 12% hypochlorite solution are mixed with 250 gal of a 2% hypochlorite solution, what is the percent strength of the solution mixture?

321. The average calcium hypochlorite use at a plant is 34 lb/day. If the chemical inventory in stock is 310 lb, how many days supply is this?

322. The flow totalizer reading a 7 a.m. on Wednesday was 43,200,000 gal and at 7 a.m. on Thursday was 44,115,670 gal. If the chlorinator setting is 18 lb for this 24-hr period, what is the chlorine dosage in mg/L?

323. A chlorine dose of 32 lb/day is required to disinfect a flow of 1,990,000 gpd. If the calcium hypochlorite to be used contains 60% available chlorine, how many pounds/day hypochlorite will be required?

324. Water from a well is disinfected by a hypochlorinator. The flow totalizer indicates that 2,666,000 gal of water were pumped during a 7-day period. The 2% sodium hypochlorite solution used to treat the well water is pumped from a 4-ft-diameter storage tank. During the 7-day period, the level in the tank dropped 3 ft, 4 in. What is the chlorine dosage in mg/L?

325. A flow of 3,350,000 gpd is to be disinfected with chlorine. If the chlorine demand is 2.5 mg/L and a chlorine residual of 0.5 mg/L is desired, what should be the chlorinator setting in lb/day?

326. If 12 gal of a 12% hypochlorite solution are mixed with 50 gal of a 1% hypochlorite solution, what is the percent strength of the solution mixture?

327. Suppose 72 lb of hypochlorite (65% available chlorine) are used in a day. If the flow rate treated is 1,880,000 gpd, what is the chlorine dosage in mg/L?

328. A hypochlorite solution (3% available chlorine) is used to disinfect a water unit. A chlorine dose of 2.6 mg/L is desired to maintain adequate chlorine residual. If the flow being treated is 400 gpm, what hypochlorite solution flow (in gpd) will be required?

329. The average daily use of sodium hypochlorite at a plant is 92 gpd. The chemical feed tank has a diameter of 3 ft and contains solution to a depth of 4 ft, 1 in. How many days supply are represented by the solution in the tank?

330. How many pounds of dry hypochlorite (65% available chlorine) must be added to 80 gal of water to make a 2% chlorine solution?

331. On average, 32 lb of chlorine are used each day at a plant. How many pounds of chlorine would be used in one week if the hour meter on the pump registers 140 hr of operation that week?

332. On average, 50 lb of chlorine are used each day at a plant. How many 150-lb chlorine cylinders will be required each month? Assume a 30-day month.

333. Express a 2.6% concentration in terms of mg/L concentration.

334. Convert 6700 mg/L to percent.

335. Express a 29% concentration in terms of mg/L.

336. Express a 22-lb/MG concentration as mg/L.

337. Convert 1.6 mg/L to lb/MG.

338. Express a 25-lb/MG concentration as mg/L.

339. Given the atomic weights for H = 1.008, Si = 28.06, and F = 19.00, calculate the percent fluoride ion present in hydrofluosilicic acid ($H_2SiF_6$).

340. Given the atom weights for Na = 22,997 and F = 19.00, calculate the percent fluoride ion present in sodium fluoride (NaF).

341. A fluoride dosage of 1.6 mg/L is desired. The flow to be treated is 980,000 gpd. How many lb/day dry sodium silicofluoride ($Na_2SiF_6$) will be required if the commercial purity of the $Na_2SiF_6$ is 98% and the percent of fluoride ion in the compound is 60.6%? Assume that the raw water contains no fluoride.

342. A fluoride dosage of 1.4 mg/L is desired. How many lb/day dry sodium silicofluoride ($Na_2SiF_6$) will be required if the flow to be treated is 1.78 MGD? The commercial purity of the sodium silicofluoride is 98% and the percent of fluoride ion in $Na_2SiF_6$ is 60.6%. Assume that the water to be treated contains no fluoride.

343. A flow of 2,880,000 gpd is to be treated with sodium silicofluoride, $Na_2SiF_6$. The raw water contains no fluoride. If the desired fluoride concentration in the water is 1.4 mg/L, what should be the chemical feed rate in lb/day? The manufacturer's data indicate that each pound of $Na_2SiF_6$ contains 0.8 lb of fluoride ion.

344. A flow of 3.08 MGD is to be treated with sodium fluoride (NaF). The raw water contains no fluoride, and the desired fluoride concentration in the finished water is 1.1 mg/L. What should be the chemical feed rate in lb/day? Each pound of NaF contains 0.45 lb of fluoride ion.

345. A flow of 810,000 gpd is to be treated with sodium fluoride (NaF). The raw water contains 0.08 mg/L fluoride and the desired fluoride level in the finished water is 1.2 mg/L. What should be the chemical feed rate in lb/day? Each pound of NaF contains 0.45 lb of fluoride ion.

346. If 9 lb of 98% pure sodium fluoride (NaF) are mixed with 55 gal of water, what is the percent strength of the solution?

347. If 20 lb of 100% pure sodium fluoride (NaF) are dissolved in 80 gal of water, what is the percent strength of the solution?

348. How many pounds of 98% pure sodium fluoride must be added to 220 gal of water to make a 1.4% solution of sodium fluoride?

349. If 11 lb of 98% pure sodium fluoride are mixed with 60 gal water, what is the percent strength of the solution?

350. How many pounds of 98% pure sodium fluoride must be added to 160 gal of water to make a 3% solution of sodium fluoride?

351. A flow of 4.23 MGD is to be treated with a 24% solution of hydrofluosilicic acid. The acid has specific gravity 1.2. If the desired fluoride level in the finished water is 1.2 mg/L, what should be the solution feed rate (in gpd)? The raw water contains no fluoride. The percent fluoride ion content of $H_2SiF_6$ is 80%.

352. A flow of 3.1 MGD nonfluoridated water is to be treated with a 22% solution of hydrofluosilicic acid ($H_2SiF_6$). The desired fluoride concentration is 1.2 mg/L. What should be the solution feed rate in gpd? The hydrofluosilicic acid weighs 9.7 lb/gal. The percent fluoride ion content of $H_2SiF_6$ is 80%.

353. A flow of 910,000 gpd is to be treated with a 2.2% solution of sodium fluoride (NaF). If the desired fluoride ion concentration is 1.8 mg/L, what should be the sodium fluoride feed rate in gpd? Sodium fluoride has a fluoride ion content of 46.10%. The water to be treated contains 0.09 mg/L fluoride ion. Assume that the solution density is 8.34 lb/gal.

354. A flow of 1,520,000 gpd nonfluoridated water is to be treated with a 2.4% solution of sodium fluoride (NaF). The desired fluoride level in the finished water is 1.6 mg/L. What should be the sodium fluoride solution feed rate in gpd? Sodium fluoride has a fluoride ion content of 45.25%. Assume that the solution density is 8.34 lb/gal.

355. The desired solution feed rate has been determined to be 80 gpd. What is this feed rate expressed as mL/min?

356. A flow of 2.78 MGD is to be treated with a 25% solution of hydrofluosilicic acid ($H_2SiF_6$) (with a fluoride content of 80%). The raw water contains no fluoride and the desired fluoride concentration is 1.0 mg/L. The hydrofluosilicic acid weighs 9.8 lb/gal. What should be the mL/min solution feed rate?

357. Suppose 40 lb/day of 98% pure sodium silicofluoride ($Na_2SiF_6$) are added to a flow of 1,520,000 gpd. The percent fluoride ion content of $Na_2SiF_6$ is 61%. What is the concentration of fluoride ions in the treated water?

358. A flow of 330,000 gpd is treated with 6 lb/day sodium fluoride (NaF). The commercial purity of the sodium fluoride is 98% and the fluoride ion content of NaF is 45.25%. Under these conditions, what is the fluoride ion dosage in mg/L?

359. A flow of 3.85 MGD nonfluoridated water is treated with a 20% solution of hydrofluosilicic acid ($H_2SiF_6$). If the solution feed rate is 32 gpd, what is the calculated fluoride ion concentration of the treated water? Assume that the acid weighs 9.8 lb/gal and the percent fluoride ion in $H_2SiF_6$ is 80%.

360. A flow of 1,920,000 gpd nonfluoridated water is treated with an 11% solution of hydrofluosilicic acid ($H_2SiF_6$). If the solution feed rate is 28 gpd, what is the calculated fluoride ion concentration of the finished water? The acid weighs 9.10 lb/gal and the percent fluoride ion in $H_2SiF_6$ is 80%.

361. A flow of 2,730,000 gpd nonfluoridated water is to be treated with a 3% saturated solution of sodium fluoride (NaF). If the solution feed rate is 110 gpd, what is the calculated fluoride ion level in the finished water? Assume that the solution weighs 8.34 lb/gal. The percent fluoride ion in NaF is 45.25%.

362. A tank contains 600 lb of 15% hydrofluosilicic acid ($H_2SiF_6$). If 2600 lb of 25% hydrofluosilicic acid are added to the tank, what is the percent strength of the solution mixture?

363. If 900 lb of 25% hydrofluosilicic acid are added to a tank containing 300 lb of 15% hydrofluosilicic acid, what will be the percent strength of the solution mixture?

364. A tank contains 400 gal of 16% hydrofluosilicic acid ($H_2SiF_6$). If a tanker truck delivers 2200 gal of 22% hydrofluosilicic acid to be added to the tank, what will be the percent strength of the solution mixture? Assume that the 22% solution weighs 9.10 lb/gal and the 16% solution weighs 9.4 lb/gal.

365. A tank contains 325 gal of 11% hydrofluosilicic acid ($H_2SiF_6$). If 1100 gal of a 20% hydrofluosilicic acid are added to the tank, what is the percent strength of the solution mixture? Assume that the 11% acid weighs 9.06 lb/gal and the 20% acid weighs 9.8 lb/gal.

366. A tank contains 220 gal of 10% hydrofluosilicic acid with a specific gravity of 1.075. If 1600 gal of 15% hydrofluosilicic acid are added to the tank, what is the percent strength of the solution mixture? Assume that the 15% acid solution weighs 9.5 lb/gal.

367. Express a 2.9% concentration in terms of mg/L.

368. Calculate the percent fluoride ion present in hydrofluosilicic acid ($H_2SiF_6$). The atomic weights are as follows: H = 1.008, Si = 28.06, and F = 19.00.

369. Express a 27-lb/MG concentration as mg/L.

370. A fluoride ion dosage of 1.6 mg/L is desired. The nonfluoridated flow to be treated is 2,111,000 gpd. How many lb/day dry 98% pure sodium silicofluoride will be required if the percent fluoride ion in the compound is 61.2%?

371. Calculate the percent fluoride ion present in sodium fluoride (NaF). The atomic weights are as follows: Na = 22.997 and F = 19.00.

372. If 80 lb of 98% pure sodium fluoride (NaF) are mixed with 600 gal of water, what is the percent strength of the solution?

373. Convert 28,000 mg/L to percent.

374. The desired solution feed rate has been determined to be 80 gpd. What is this feed rate expressed in mL/min?

375. A fluoride dosage of 1.5 mg/L is desired. How many lb/day of 98% pure dry sodium fluoride (NaF) will be required if the flow to be treated is 2.45 MGD? The percent fluoride ion in NaF is 45.25%.

376. How many pounds of 98% pure sodium fluoride must be added to 600 gal of water to make a 3% solution of sodium fluoride?

377. A flow of 4.11 MGD nonfluoridated water is to be treated with a 21% solution of hydrofluosilicic acid ($H_2SiF_6$). The acid has a specific gravity of 1.3. If the desired fluoride level in the finished water is 1.4 mg/L, what should be the solution feed rate in gpd? The percent fluoride ion content of the acid is 80%.

378. If 30 lb of 98% pure sodium fluoride are mixed with 140 gal water, what is the percent strength of the solution?

379. A flow of 1,880,000 gpd is to be treated with sodium fluoride (NaF) containing 0.44 lb of fluoride ions. The raw water contains 0.09 mg/L and the desired fluoride level in the finished water is 1.4 mg/L. What should be the chemical feed rate in lb/day?

380. A flow of 2.8 MGD nonfluoridated water is to be treated with a 20% solution of hydrofluosilicic acid ($H_2SiF_6$). The desired fluoride concentration is 1.3 mg/L. What should be the solution feed rate in gpd? The hydrofluosilicic acid weighs 9.8 lb/gal, and the percent fluoride ion content of acid is 80%.

381. A tank contains 500 lb of 15% hydrofluosilicic acid. If 1600 lb of 20% hydrofluosilicic acid are added to the tank, what is the percent strength of the solution mixture?

382. Suppose 41 lb/day of sodium silicofluoride are added to a flow of 1,100,000 gpd. The commercial purity of sodium silicofluoride is 98% and the percent fluoride ion content of acid is 61%. What is the concentration of fluoride in the treated water?

383. A flow of 1400-gpm nonfluoridated raw water is treated with an 11% solution of hydrofluosilicic acid. If the solution feed rate is 40 gpd, what is the calculated fluoride ion concentration of the finished water? The acid weighs 9.14 lb/gal, and the percent fluoride ion in the acid is 80%.

384. A tank contains 235 gal of 10% hydrofluosilicic acid. If 600 gal of a 20% hydrofluosilicic acid are added to the tank, what is the percent strength of the solution mixture? Assume that the 10% acid weighs 9.14 lb/gal and the 20% acid weighs 9.8 lb/gal.

385. A flow of 2.88 MGD is to be treated with a 20% solution of hydrofluosilicic acid. The raw water contains no fluoride and the desired fluoride concentration is 1.1 mg/L. The acid weighs 9.8 lb/gal. What should be the mL/min solution feed rate? The percent fluoride content of acid is 80%.

386. A tank contains 131 gal of 9% hydrofluosilicic acid with a specific gravity of 1.115. If 900 gal of 15% hydrofluosilicic acid are added to the tank, what is the percent strength of the solution mixture? Assume that the 15% acid solution weighs 9.4 lb/gal.

387. A flow of 2,900,000 gpd is to be treated with a 5% saturated solution of sodium fluoride (NaF). If the solution feed rate is 120 gpd, what is the calculated fluoride ion level in the finished water? Assume that the solution weighs 8.34 lb/gal. The percent fluoride ion in the acid is 45.25%. The raw water contains 0.2 mg/L fluoride.

388. The calcium content of a water sample is 39 mg/L. What is this calcium hardness expressed as $CaCO_3$? The equivalent weight of calcium is 20.04, and the equivalent weight of $CaCO_3$ is 50.045.

389. The magnesium content of a water unit is 33 mg/L. What is this magnesium hardness expressed as $CaCO_3$? The equivalent weight of magnesium is 12.16, and the equivalent weight of $CaCO_3$ is 50.045.

390. A water unit contains 18 mg/L calcium. What is this calcium hardness expressed as $CaCO_3$? The equivalent weight of calcium is 20.04, and the equivalent weight of $CaCO_3$ is 50.045.

391. A water unit has a calcium concentration of 75 mg/L as $CaCO_3$ and a magnesium concentration of 91 mg/L as $CaCO_3$. What is the total hardness (as $CaCO_3$) of the sample?

392. Determine the total hardness as $CaCO_3$ of a water unit that has calcium content of 30 mg/L and magnesium content of 10 mg/L. The equivalent weight of calcium is 20.04, the equivalent weight of magnesium is 12.16, and the equivalent weight of $CaCO_3$ is 50.045.

393. Determine the total hardness as $CaCO_3$ of a water unit that has calcium content of 21 mg/L and magnesium content of 15 mg/L. The equivalent weight of calcium is 20.04, the equivalent weight of magnesium is 12.16, and the equivalent weight of $CaCO_3$ is 50.045.

394. A sample of water contains 125-mg/L alkalinity as $CaCO_3$. If the total hardness of the water is 121 mg/L as $CaCO_3$, what are the carbonate hardness and noncarbonate hardness in mg/L as $CaCO_3$?

395. The alkalinity of a water unit is 105 mg/L as $CaCO_3$. If the total hardness of the water is 122 mg/L as $CaCO_3$, what are the carbonate hardness and noncarbonate hardness in mg/L as $CaCO_3$?

396. A water unit has an alkalinity of 91 mg/L as $CaCO_3$ and a total hardness of 116 mg/L. What are the carbonate hardness and noncarbonate hardness of the water?

397. A water sample contains 112 mg/L alkalinity as $CaCO_3$ and 99 mg/L total hardness as $CaCO_3$. What are the carbonate hardness and noncarbonate hardness of this water?

398. The alkalinity of a water unit is 103 mg/L as $CaCO_3$. If the total hardness of the water is 121 mg/L as $CaCO_3$, what are the carbonate hardness and noncarbonate hardness in mg/L as $CaCO_3$?

399. A 100-mL water sample is tested for phenolphthalein alkalinity. If 2 mL titrant are used to titrate to pH 8.3 and the sulfuric acid solution has a normality of 0.02 *N*, what is the phenolphthalein alkalinity of the water in mg/L as $CaCO_3$?

400. A 100-mL water sample is tested for phenolphthalein alkalinity. If 1.4 mL titrant are used to titrate to pH 8.3 and the normality of the sulfuric acid solution is 0.02 *N*, what is the phenolphthalein alkalinity of the water in mg/L as $CaCO_3$?

401. A 100-mL sample of water is tested for alkalinity. The normality of the sulfuric acid used for titrating is 0.02 *N*. If 0.3 mL titrant is used to titrate to pH 8.3, and 6.7 titrant is used to titrate to pH 4.4, what are the phenolphthalein and total alkalinity of the sample?

402. A 100-mL sample of water is tested for phenolphthalein and total alkalinity, and 0 mL titrant is used to titrate to pH 8.3 and 6.9 mL titrant is used to titrate to pH 4.4. The normality of the acid used for titrating is 0.02 *N*. What are the phenolphthalein and total alkalinity of the sample in mg/L as $CaCO_3$?

403. A 100-mL sample of water is tested for alkalinity. The normality of the sulfuric acid used for titrating is 0.02 *N*. If 0.5-mL titrant is used to titrate to pH 8.3, and 5.7-mL titrant is used to titrate to pH 4.6, what are the phenolphthalein and total alkalinity of the sample? (Refer to Table 8.1.)

404. A water sample is tested for phenolphthalein and total alkalinity. If the phenolphthalein alkalinity is 8 mg/L as $CaCO_3$ and the total alkalinity is 51 mg/L as $CaCO_3$, what are the bicarbonate, carbonate, and hydroxide alkalinity of the water?

405. A water sample is found to have a phenolphthalein alkalinity of 0 mg/L and a total alkalinity of 67 mg/L. What are the bicarbonate, carbonate, and hydroxide alkalinity of the water?

406. The phenolphthalein alkalinity of a water sample is 12 mg/L as $CaCO_3$ and the total alkalinity is 23 mg/L as $CaCO_3$. What are the bicarbonate, carbonate, and hydroxide alkalinity of the water?

407. Alkalinity titrations on a 100-mL water sample resulted in the following: 1.3 mL titrant were used to titrate to pH 8.3, and 5.3 mL titrant were used to titrate to pH 4.6. The normality of the sulfuric acid was 0.02 *N*. What are the phenolphthalein, total, bicarbonate, carbonate, and hydroxide alkalinity of the water?

408 Alkalinity titrations on a 100-mL water sample resulted in the following: 1.5 mL titrant were used to titrate to pH 8.3, and 2.9 mL titrant were used to titrate to pH 4.5. The normality of the sulfuric acid was 0.02 *N*. What are the phenolphthalein, total, bicarbonate, carbonate, and hydroxide alkalinity of the water?

409. Assuming 15% excess lime, a water sample has a carbon dioxide content of 8 mg/L, as $CO_2$, total alkalinity of 130 mg/L as $CaCO_3$, and magnesium content of 22 mg/L as $Mg^{2+}$. Approximately how much quicklime (CaO; 90% purity) will be required for softening?

410. Assuming 15% excess lime, the characteristics of a water unit are as follows: 5 mg/L $CO_2$, 164 mg/L total alkalinity as $CaCO_3$, and 17 mg/L magnesium as $Mg^{2+}$. What is the estimated hydrated lime ($Ca(OH)_2$; 90% pure) dosage in mg/L required for softening?

411. Assuming 15% excess lime, a water sample has the following characteristics: 6 mg/L $CO_2$, 110 mg/L total alkalinity as $CaCO_3$, and 12 mg/L magnesium as $Mg^{2+}$. What is the estimated hydrated lime ($Ca(OH)_2$; 90% purity) dosage in mg/L required for softening?

412. A water sample has a carbon dioxide content of 9 mg/L as $CO_2$, total alkalinity of 180 mg/L as $CaCO_3$, and magnesium content of 18 mg/L as $MG^{2+}$. About how much quicklime (CaO; 90% purity) will be required for softening?

413. A water unit has a total hardness of 260 mg/L as $CaCO_3$ and a total alkalinity of 169 mg/L. What soda ash dosage (mg/L) will be required to remove the noncarbonate hardness?

414. The alkalinity of a water unit is 111 mg/L as $CaCO_3$, and the total hardness is 240 mg/L as $CaCO_3$. What soda ash dosage (in mg/L) is required to remove the noncarbonate hardness?

415. A water sample has a total hardness of 264 mg/L as $CaCO_3$ and a total alkalinity of 170 mg/L. What soda ash dosage (in mg/L) will be required to remove the noncarbonate hardness?

416. Calculate the soda ash required (in mg/L) to soften a water unit if the water has a total hardness of 228 mg/L and a total alkalinity of 108 mg/L.

417. The *A*, *B*, *C*, and *D* factors of the excess lime equation have been calculated as follows: $A = 8$ mg/L, $B = 130$ mg/L, $C = 0$, and $D = 66$ mg/L. If the residual magnesium is 4 mg/L, what is the carbon dioxide dosage (in mg/L) required for recarbonation?

418. The *A*, *B*, *C*, and *D* factors of the excess lime equation have been calculated as follows: $A = 8$ mg/L, $B = 90$ mg/L, $C = 7$, and $D = 108$ mg/L. If the residual magnesium is 3 mg/L, what carbon dioxide dosage (in mg/L) would be required for recarbonation?

419. The *A*, *B*, *C*, and *D* factors of the excess lime equation were determined to be as follows: *A* = 7 mg/L, *B* = 109 mg/L, *C* = 3, and *D* = 52 mg/L. If the magnesium residual is 5 mg/L, what is the carbon dioxide dosage (in mg/L) required for recarbonation?

420. The *A*, *B*, *C*, and *D* factors of the excess lime equation were determined to be as follows: *A* = 6 mg/L, *B* = 112 mg/L, *C* = 6, and *D* = 45 mg/L. If the residual magnesium is 4 mg/L, what carbon dioxide dosage (in mg/L) would be required for recarbonation?

421. Jar tests indicate that the optimum lime dosage is 200 mg/L. If the flow to be treated is 2.47 MGD, what should be the chemical feeder setting in lb/day?

422. The optimum lime dosage for a water unit has been determined to be 180 mg/L. If the flow to be treated is 3,120,000 gpd, what should be the chemical feeder setting in lb/day and lb/min?

423. A soda ash dosage of 60 mg/L is required to remove noncarbonate hardness. What should be the lb/hr chemical feeder setting if the flow rate to be treated is 4.20 MGD?

424. What should be the chemical feeder settings (in lb/day and lb/min) if the optimum lime dosage has been determined to be 130 mg/L and the flow to be treated is 1,850,000 gpd?

425. If 40 mg/L soda ash are required to remove noncarbonate hardness from a water process, what should be the chemical feeder settings (in lb/hr and lb/min) if the flow to be treated is 3,110,000 gpd?

426. The total hardness of a water unit is 211 mg/L. What is this hardness expressed as grains per gallon?

427. The total hardness of a water sample is 12.3 mg/L. What is this concentration expressed as mg/L?

428. The total hardness of a water unit is reported as 240 mg/L. What is the hardness expressed as grains per gallon?

429. A hardness of 14 gr/gal is equivalent to how many mg/L?

430 The hardness removal capacity of an ion exchange resin is 25,000 gr/ft$^3$. If the softener contains 105 ft$^3$ of resin, what is the exchange capacity of the softener in grains?

431. An ion exchange water softener has a diameter of 6 ft. The depth of the resin is 4.2 ft. If the resin has a removal capacity of 25 Kgr/ft$^3$, what is the exchange capacity of the softener in grains?

432. The hardness removal capacity of an exchange resin is 20 Kgr/ft$^3$. If the softener contains 260 ft$^3$ of resin, what is the exchange capacity of the softener in grains?

433. An ion exchange water softener has a diameter of 8 ft. The depth of the resin is 5 ft. If the resin has a removal capacity of 22 Kgr/ft$^3$, what is the exchange capacity of the softener in grains?

434. An ion exchange softener has an exchange capacity of 2,210,000 grains. If the hardness of the water to be treated is 18.1 gr/gal, how many gallons of water can be treated before regeneration of the resin is required?

435. The exchange capacity of an ion exchange softener is 4,200,000 grains. If the hardness of the water to be treated is 16.0 gr/gal, how many gallons of water can be treated before regeneration of the resin is required?

436. An ion exchange softener has an exchange capacity of 3,650,000 grains. If the hardness of the water to be treated is 270 mg/L, how many gallons of water can be treated before regeneration of the resin is required?

437. The hardness removal capacity of an ion exchange resin is 21 Kgr/ft$^3$. The softener contains 165 ft$^3$ of resin. If the water to be treated contains 14.6 gr/gal hardness, how many gallons of water can be treated before regeneration of the resin is required?

438. The hardness removal capacity of an ion exchange resin is 22,000 gr/ft$^3$. The softener has a diameter of 3 ft and a depth of resin of 2.6 ft. If the water to be treated contains 12.8 gr/gal hardness, how many gallons of water can be treated before regeneration of the resin is required?

439. An ion exchange softener can treat 575,000 gal of water before regeneration is required. If the flow rate treated is 25,200 gph, how many hours of operation remain before regeneration is required?

440. An ion exchange softener can treat 766,000 gal of water before regeneration of the resin is required. If the water is to be treated at a rate of 26,000 gph, how many hours of operation remain until regeneration is required?

441. Suppose 348,000 gal of water can be treated by an ion exchange water softener before regeneration of the resin is required. If the flow rate to be treated is 230 gpm, what is the operating time (in hr) until regeneration of the resin will be required?

442. The exchange capacity of an ion exchange softener is 3,120,000 grains. The water to be treated contains 14 gr/gal total hardness. If the flow rate to be treated is 200 gpm, how many hours of operation remain until regeneration of the resin will be required?

443. The exchange capacity of an ion exchange softener is 3,820,000 grains. The water to be treated contains 11.6 gr/gal total hardness. If the flow rate to be treated is 290,000 gpd, how many hours of operation are there until regeneration of the resin will be required?

444. An ion exchange softener will remove 2300 Kgr hardness from the water until the resin must be regenerated. If 0.5 lb salt is required for each kilograin removed, how may pounds of salt will be required for preparing the brine to be used in resin regeneration?

445. If 1330 Kgr hardness are removed by an ion exchange softener before the resin must be regenerated and 0.4 lb salt is required for each kilograin removed, how many pounds of salt will be required for preparing the brine to be used in resin regeneration?

446. Suppose 410 lb salt are required in making up a brine solution for regeneration. If the brine solution is to be a 13% solution of salt, how many gallons of brine will be required for regeneration of the softener? (Use Table 8.2 to determine the lb salt per gal brine for a 13% solution.)

447. Suppose 420 lb salt are required to regenerate an ion exchange softener. If the brine solution is to be a 14% brine solution, how many gallons brine will be required? (Use Table 8.2 to determine the lb salt per gal brine for a 14% brine solution.)

448. An ion exchange softener removes 1310 Kgr hardness from the water before the resin must be regenerated, and 0.5 lb salt is required for each kilograin hardness removed. If the brine solution is to be a 12% brine solution, how many gallons of brine will be required for regeneration of the softener? (Use Table 8.2 to determine the lb salt per gal brine for a 12% brine solution.)

449. The calcium content of a water sample is 44 mg/L. What is this calcium hardness expressed as $CaCO_3$? The equivalent weight of calcium is 20.04, and the equivalent weight of $CaCO_3$ is 50.045.

450. A 100-mL sample of water is tested for phenolphthalein alkalinity. If 1.8 mL titrant are used to titrate to pH 8.3 and the normality of the sulfuric acid solution is 0.02 *N*, what is the phenolphthalein alkalinity of the water in mg/L as $CaCO_3$?

451. The magnesium content of a water sample is 31 mg/L. What is this magnesium hardness expressed as $CaCO_3$? The equivalent weight of magnesium is 12.16, and the equivalent weight of $CaCO_3$ is 50.045.

452. How many milliequivalents of calcium do 24 mg of magnesium equal?

453. The characteristics of a water unit are as follows: 8 mg/L $CO_2$, 118 mg/L total alkalinity as $CaCO_3$, and 12 mg/L magnesium as $Mg^{2+}$. What is the estimated hydrated lime ($Ca(OH)_2$; 90% pure) dosage in mg/L required for softening?

454. Determine the total hardness of $CaCO_3$ of a water unit that has calcium content of 31 mg/L and magnesium content of 11 mg/L. The equivalent weight of calcium is 20.04, the equivalent weight of magnesium is 12.16, and the equivalent weight of $CaCO_3$ is 50.045.

455. A sample of water contains 112 mg/L alkalinity as $CaCO_3$ and 101 mg/L total hardness of $CaCO_3$. What are the carbonate hardness and noncarbonate hardness of this water?

456. A water sample has a carbon dioxide content of 5 mg/L as $CO_2$, total alkalinity of 156 mg/L as $CaCO_3$, and magnesium content of 11 mg/L as $MG^{2+}$. Approximately how much quicklime (CaO; 90% purity) will be required for softening? (Assume 15% excess lime.)

# Appendix: Solutions to Chapter 9 Problems

1. 98 ft – 91 ft = 7 ft drawdown
2. 125 ft – 110 ft = 15 ft drawdown
3. 161 ft – 144 ft = 17 ft drawdown
4. 3.7 psi × 2.31 ft/psi = 8.5 ft water depth in sounding line
   112 ft – 8.5 ft = 103.5 ft
   103.5 ft – 86 ft = 17.5 ft drawdown
5. 4.6 psi × 2.31 ft/psi = 10.6 water depth in sounding line
   150 ft – 10.6 ft = 139.4 ft
   171 ft – 139.4 ft = 31.4 ft drawdown
6. 300 ÷ 20 = 15 gpm per ft of drawdown
7. 420 gal ÷ 5 min. = 84 gpm
8. 810 gal ÷ 5 min. = 162 gpm
9. 856 gal ÷ 5 min. = 171 gpm
   171 gpm × 60 min/hr = 10,260 gph
10. $$\frac{0.785 \times 1 \text{ ft} \times 1 \text{ ft} \times 12 \text{ ft} \times 7.48 \text{ gal/ft}^3 \times 12 \text{ round trips}}{5 \text{ minutes}} = 169 \text{ gpm}$$
11. 750 gal ÷ 5 min = 150 gpm
    150 gpm × 60 min/hr = 9,000 gph
    9000 gph × 10 hr/day = 90,000 gpd
12. 200 gpm ÷ 28 ft = 7.1 gpm/ft
13. 620 gpm ÷ 21 ft = 29.5 gpm/ft
14. 1100 gpm ÷ 41.3 ft = 26.6 gpm/ft
15. $$\frac{x \text{ gpm}}{42.8 \text{ ft}} = 33.4 \text{ fpm/ft}$$
    $$x = 33.4 \text{ fpm/ft} \times 42.8 \text{ ft} = 2540 \text{ gpm}$$
16. 0.785 × 0.5 ft × 0.5 ft × 140 ft × 7.48 gal/ft$^3$ = 206 gal
    40 mg/L × 0.000206 MG × 8.34 lb/gal = 0.07 lb
17. 0.785 × 1 ft × 1 ft × 109 ft × 7.48 gal/ft$^3$ = 640 gal
    40 mg/L × 0.000640 MG × 8.34 lb/gal = 0.21 lb
18. 0.785 × 1 ft × 1 ft × 109 ft × 7.48 gal/ft$^3$ = 633 gal
    0.785 × 0.67 ft × 0.67 ft × 40 ft × 7.48 gal/ft$^3$ = 105 gal
    633 + 105 gal = 738 gal
    110 mg/L × 0.000738 MG × 8.34 lb/gal = 0.68 lb
19. $$x \text{ mg/L} \times 0.000540 \text{ MG} \times 8.34 \text{ lb/gal} = 0.48 \text{ lb}$$
    $$x = \frac{0.48 \text{ lb}}{0.000540 \text{ MG} \times 8.34 \text{ lb/gal}} = 107 \text{ mg/L}$$

20. $\frac{0.09 \text{ lb}}{5.25/100} = 1.5 \text{ lb}$

$\frac{1.5 \text{ lb}}{8.34 \text{ lb/gal}} = 0.18 \text{ gal}$

$0.18 \text{ gal} \times 128 \text{ fluid oz./gal} = 23 \text{ fluid oz.}$

21. $0.785 \times 0.5 \text{ ft} \times 0.5 \text{ ft} \times 120 \text{ ft} \times 7.48 \text{ gal/ft}^3 = 176 \text{ gal}$

$\frac{50 \text{ mg/L} \times 0.000176 \text{ MG} \times 8.34 \text{ lb/gal}}{65/100} = 0.1 \text{ lb}$

$0.1 \text{ lb} \times 16 \text{ oz./lb} = 1.6 \text{ oz. calium hypochlorite}$

22. $0.785 \times 1.5 \text{ ft} \times 1.5 \text{ ft} \times 105 \text{ ft} \times 7.48 \text{ gal/ft}^3 = 1387 \text{ gal}$

$\frac{100 \text{ mg/L} \times 0.001387 \text{ MG} \times 8.34 \text{ lb/gal}}{25/100} = 4.6 \text{ lb chloride of lime}$

23. $\frac{60 \text{ mg/L} \times 0.000240 \text{ MG} \times 8.34 \text{ lb/gal}}{5.25/100} = 2.3 \text{ lb}$

$\frac{2.3 \text{ lb}}{8.34 \text{ lb/gal}} = 0.3 \text{ gal}$

$0.3 \text{ gal} \times 128 \text{ fluid oz./gal} = 38.4 \text{ fluid oz. sodium hypochlorite}$

24. 4.0 psi × 2.31 ft/psi = 9.2 ft
25. (94 ft + 24 ft) + (3.6 psi × 2.31 ft/psi) = 118 ft + 8.3 ft = 126.3 ft
26. 4.2 psi × 2.31 ft/psi = 9.7 ft
    180 ft + 9.7 ft = 189.7 ft
    whp = (189.7 ft × 800 gpm) ÷ 3960 = 38.3
27. 4.4 psi × 2.31 ft/psi = 10.2 ft
    Field head = 200 ft + 10.2 ft = 210.2 ft
    whp = (210.2 ft × 1000 gpm) ÷ 3960 = 53
28. $\text{Bowl bhp} = \frac{184 \text{ ft} \times 700 \text{ gpm}}{3960 \times (83/100)} = 39$
29. Shaft friction loss = 0.67 hp

$\frac{0.67 \text{ hp}}{100 \text{ ft}} = 181 \text{ ft} = 1.2 \text{ hp loss}$

$\text{Field bhp} = 59.5 \text{ bhp} + 1.2 \text{ hp} = 60.7 \text{ hp}$

30. $\text{mhp} = \frac{\text{Total bhp}}{\text{Motor efficiency}/100} = \frac{58.3 \text{ bhp} + 0.5 \text{ hp}}{90/100} = 65.3 \text{ hp}$
31. (45 hp/56.4 hp) × 100 = 80%
32. $\frac{55.7 \text{ hp}}{90/100} = 62 \text{ hp input;} \quad \frac{43.5 \text{ hp}}{62 \text{ hp}} = 70\%$
33. 400 ft × 110 ft × 14 ft × 7.48 gal/ft$^3$ = 4,607,680 gal

34. 400 ft × 110 ft × 30 ft × 0.4 average depth × 7.48 gal/ft³ = 3,949,440 gal

35. $\frac{200 \text{ ft} \times 80 \text{ ft} \times 12 \text{ ft}}{43{,}560 \text{ ft}^3/\text{ac-ft}} = 4.4 \text{ ac-ft}$

36. $\frac{320 \text{ ft} \times 170 \text{ ft} \times 16 \text{ ft} \times 0.4}{43{,}560 \text{ ft}^3/\text{ac-ft}} = 8.0 \text{ ac-ft}$

37. $\frac{0.5 \text{ mg/L} \times 20 \text{ MG} \times 8.34 \text{ lb/gal}}{25/100} = 334 \text{ lb copper sulfate}$

38. 62 ac-ft × 43,560 ft³/ac-ft × 7.48 gal/ft³ = 20,201,385 gal

$$\frac{0.5 \text{ mg/L} \times 20.2 \text{ MG} \times 8.34 \text{ lb/gal}}{25/100} = 337 \text{ lb copper sulfate}$$

39. $\frac{1.1 \text{ lb } CuSO_4}{1 \text{ ac-ft}} \times 38 \text{ ac-ft} = 41.8 \text{ lb copper sulfate}$

40. $\frac{250 \text{ ft} \times 75 \text{ ft} \times 10 \text{ ft}}{43{,}560 \text{ ft}^3/\text{ac-ft}} = 4.3 \text{ ac-ft}$

$$\frac{0.8 \text{ lb } CuSO_4}{1 \text{ ac-ft}} \times 4.3 \text{ ac-ft} = 3.14 \text{ lb copper sulfate}$$

41. $\frac{500 \text{ ft} \times 100 \text{ ft}}{43{,}560 \text{ ft}^2/\text{ac}} = 1.1 \text{ ac}$

$$\frac{5.1 \text{ lb } CuSO_4}{1 \text{ ac}} \times 1.1 \text{ ac} = 5.9 \text{ lb copper sulfate}$$

42. 131.9 ft – 93.5 ft = 38.4 ft

43. 707 gal ÷ 5 minutes = 141 gpm; 141 gpm × 60 min/hr = 8460 gph

44. $\frac{0.785 \times 1 \text{ ft} \times 1 \text{ ft} \times 12 \text{ ft} \times 7.48 \text{ gal/ft}^3 \times 8 \text{ round trips}}{5 \text{ gpm}} = 113 \text{ gpm}$

45. 3.5 psi × 2.31 ft/psi = 8.1 water depth in sounding line
167 ft – 8.1 ft = 158.9 ft pumping water level
Drawdown = 158.9 ft – 141 ft = 17.9 ft

46. 610 gpm ÷ 28 ft = 21.8 gpm/ft

47. 0.785 × 0.5 ft × 0.5 ft × 150 ft × 7.48 gal/ft³ = 220 gal
Chlorine required = 55 mg/L × 0.000220 MG × 8.34 lb/gal = 0.10 lb

48. 780 gal ÷ 5 min = 156 gpm; 156 gal/min × 60 min/hr × 8 hr/day = 74,880 gpd

49. $x \text{ mg/L} \times 0.000610 \text{ MG} \times 8.34 \text{ lb/gal} = 0.47 \text{ lb}$

$$x = \frac{0.47 \text{ lb}}{0.000610 \text{ MG} \times 8.34 \text{ lb/gal}} = 92.3 \text{ mg/L}$$

50. 0.785 × 1 ft × 1 ft × 89 ft × 7.48 gal/ft³ = 523 gal
0.785 × 0.67 ft × 0.67 ft × 45 ft × 7.48 gal/ft³ = 119 gal
523 gal + 119 gal = 642 gal
100 mg/L × 0.000642 MG × 8.34 lb/gal = 0.54 lb chlorine

51. $\frac{0.3\text{ lb}}{5.25/100} = 5.7\text{ lb}; \quad \frac{5.7\text{ lb}}{8.34\text{ lb/gal}} = 0.68\text{ lb}$

$0.68\text{ gal} \times 128\text{ fluid oz./gal} = 87\text{ fluid oz.}$

52. 4.5 psi × 2.31 psi = 10.4 ft
53. 3.6 psi × 2.31 ft/psi = 8.3 ft
95 ft + 25 ft + 8.3 ft = 128.3 ft
54. 191 ft + (4.1 psi × 2.31 ft/psi) = 191 ft + 9.5 ft = 200.5 ft
whp = (200.5 ft × 850 gpm) ÷ 3960 gal/min/ft = 43 hp

55. $\frac{175\text{ ft} \times 800\text{ gpm}}{3960\text{ gal/min/ft} \times 0.80} = 44.2\text{ hp}$

56. $\frac{47.8\text{ hp} + 0.8\text{ hp}}{0.90} = 54\text{ hp}$

57. $\frac{45.6\text{ hp}}{57.4\text{ hp}} \times 100 = 79.8\%$

58. $\frac{54.7\text{ hp}}{0.90} = 61\text{ hp}$

$\text{Overall efficiency } (\%) = \frac{44.6\text{ hp}}{61\text{ hp}} = 73\%$

59. 53 ac-ft × 43,560 $ft^3$/ac-ft = 2,308,680 $ft^3$
2,308,680 $ft^3$ × 7.48 gal/$ft^3$ = 17,268,926 gal
(0.5 mg/L × 17.2 MG × 8.34 lb/gal) ÷ 0.25 = 287 lb copper sulfate
60. (440 ft × 140 ft) ÷ 43,560 $ft^2$/ac = 1.4 ac
5.5 lb × 1.4 ac = 7.7 lb/ac
61. 4 ft × 5 ft × 3 ft × 7.48 gal/$ft^3$ = 449 gal
62. 50 ft × 20 ft × 8 ft × 7.48 gal/$ft^3$ = 59,840 gal
63. 40 ft × 16 ft × 8 ft × 7.48 gal/$ft^3$ = 38,298 gal
64. 42 in. ÷ (12 in./ft) = 3.5 ft
5 ft × 5 ft × 3.5 ft × 7.48 gal/$ft^3$ = 655 gal
65. 2 in. ÷ 12 in./ft = 0.17 ft
40 ft × 25 ft × 9.17 ft × 7.48 gal/$ft^3$ = 68,592 gal
66. 3,625,000 gpd ÷ 1440 min/day = 2517 gpm
(60 ft × 25 ft × 9 ft × 7.48 gal/$ft^3$) ÷ 2517 gpm = 40.1 min
67. 2,800,000 gpd ÷ 1440 min/day = 1944 gpm

$\text{Detention time} = \frac{50\text{ ft} \times 20\text{ ft} \times 8\text{ ft} \times 7.48\text{ gal/ft}^3}{1944\text{ gpm}} = 30.8\text{ min}$

68. 9,000,000 gpd ÷ (1440 min/day × 60 sec/min) = 104.2 gps

$\text{Detention time} = \frac{6\text{ ft} \times 5\text{ ft} \times 5\text{ ft} \times 7.48\text{ gal/ft}^3}{104.2\text{ gps}} = 10.8\text{ sec}$

69. 2,250,000 gpd ÷ 1440 min/day = 1563 gpm

$$\text{Detention time} = \frac{50\text{ ft} \times 20\text{ ft} \times 10\text{ ft} \times 7.48\text{ gal/ft}^3}{1563\text{ gpm}} = 47.9\text{ min}$$

70. 3,250,000 gpd ÷ (1440 min/day × 60 sec/min) = 37.6 gps

$$\text{Detention time} = \frac{4\text{ ft} \times 4\text{ ft} \times 3.5\text{ ft} \times 7.48\text{ gal/ft}^3}{37.6\text{ gps}} = 11.1\text{ sec}$$

71. 10 mg/L × 3.45 MGD × 8.34 lb/gal = 288 lb/day
72. 12 mg/L × 1.660 MGD × 8.34 lb/gal = 166 lb/day
73. 10 mg/L × 2.66 MGD × 8.34 lb/gal = 222 lb/day
74. 9 mg/L × 0.94 MGD × 8.34 lb/gal = 71 lb/day
75. 12 mg/L × 4.10 MGD × 8.34 lb/gal = 410 lb/day
76. 7 mg/L × 1.66 MGD × 8.34 lb/gal = 97 lb/day
    97 lb/day ÷ 5.24 lb/gal = 18.5 gpd
77. $12\text{ mg/L} \times 3.43\text{ MGD} \times 8.34\text{ lb/gal} = 550{,}000\text{ mg/L} \times x\text{ MGD} \times 8.34\text{ lb/gal}$

$$\frac{12\text{ mg/L} \times 3.43\text{ MGD} \times 8.34\text{ lb/gal}}{550{,}000\text{ mg/L} \times 8.34\text{ lb/gal}} = x\text{ MGD}$$

$$0.0000748\text{ MGD} = x$$

78. 10 mg/L × 4.13 MGD × 8.34 lb/gal = 344 lb/day
    344 lb/day ÷ 5.40 lb/gal = 64 gpd
79. $11\text{ mg/L} \times 0.88\text{ MGD} \times 8.34\text{ lb/gal} = 550{,}000\text{ mg/L} \times x\text{ MGD} \times 8.34\text{ lb/gal}$

$$\frac{11\text{ mg/L} \times 0.88\text{ MGD} \times 8.34\text{ lb/gal}}{550{,}000\text{ mg/L} \times 8.34\text{ lb/gal}} = x\text{ MGD}$$

$$0.0000176\text{ MGD} = x$$

80. $\frac{640\text{ mg}}{1\text{ mL}} \times \frac{1000}{1000} = \frac{640{,}000\text{ mg}}{1000\text{ mL}} = 640{,}000\text{ mg/L}$

$$\frac{10\text{ mg/L} \times 1.85\text{ MGD} \times 8.34\text{ lb/gal}}{640{,}000\text{ mg/L} \times 8.34\text{ lb/gal}} = 0.0000289\text{ MGD} = 28.9\text{ gpd}$$

81. (40 gpd × 3785 mL/gal) ÷ 1440 min/day = 105 mL/min
82. (34.2 gpd × 3785 mL/gal) ÷ 1440 min/day = 90 mL/min
83. $10\text{ mg/L} \times 2.88\text{ MGD} \times 8.34\text{ lb/gal} = 550{,}000\text{ mg/L} \times x\text{ MGD} \times 8.34\text{ lb/gal}$

$$\frac{10\text{ mg/L} \times 2.88\text{ MGD} \times 8.34\text{ lb/gal}}{550{,}000\text{ mg/L} \times 8.34\text{ lb/gal}} = x\text{ MGD}$$

$$0.0000523\text{ MGD (52.4 gpd)} = x$$

(52.4 gpd × 3785 mL/gal) ÷ 1440 min/day = 138 mL/min

84. $6 \text{ mg/L} \times 2.82 \text{ MGD} \times 8.34 \text{ lb/gal} = 550{,}000 \text{ mg/L} \times x \text{ MGD} \times 8.34 \text{ lb/gal}$

$$\frac{6 \text{ mg/L} \times 2.82 \text{ MGD} \times 8.34 \text{ lb/gal}}{550{,}000 \text{ mg/L} \times 8.34 \text{ lb/gal}} = x \text{ MGD}$$

$$0.0000307 \text{ MGD (30.7 gpd)} = x$$

(30.7 gpd × 3785 mL/gal) ÷ 1440 min/day = 80.7 mL/min

85. $$\frac{10 \text{ mg/L} \times 3.45 \text{ MGD} \times 8.34 \text{ lb/gal}}{5.40 \text{ lb/gal}} = 53.3 \text{ mL/min}$$

(53.3 gpd × 3785 mL/gal) ÷ 1440 min/day = 141 mL/min

86. $140 \text{ g} = 0.0022 \text{ lb/g} = 0.31 \text{ lb}$

$$\frac{0.31 \text{ lb}}{(16 \text{ gal} \times 8.34 \text{ lb/gal}) + 0.31 \text{ lb}} \times 100 = 0.23\%$$

87. 22 oz. ÷ 16 oz./lb = 1.38 lb

$$\frac{1.38 \text{ lb}}{(24 \text{ gal} \times 8.34 \text{ lb/gal}) + 1.38 \text{ lb}} \times 100 = 0.68\%$$

88. $$\frac{2.1 \text{ lb} \times 100}{(x \text{ gal} \times 8.34 \text{ lb/gal}) + 2.1 \text{ lb}} = 0.80\%$$

$$\frac{210}{8.34x + 2.1} = 0.8$$

$$210 = 0.8 \times (8.34x + 2.1)$$

$$210 = 6.7x + 1.7$$

$$208.3 = 6.7x$$

$$31 \text{ gal} = x$$

89. $$0.11x = 0.005 \times 160 \text{ lb}$$

$$x = \frac{0.005 \times 160 \text{ lb}}{0.11} = 7.27 \text{ lb}$$

90. $$1.3 \times 8.34 \text{ lb/gal} = 10.8 \text{ lb/gal}$$

$$0.08 \times x \text{ gal} \times 10.8 \text{ lb/gal} = 0.002 \times 50 \text{ gal} \times 8.34 \text{ lb/gal}$$

$$x = \frac{0.002 \times 50 \text{ gal} \times 8.34 \text{ lb/gal}}{0.08 \times 10.8 \text{ lb/gal}} = 0.86 \text{ gal}$$

91. $0.11 \times x \text{ gal} \times 10.1 \text{ lb/gal} = 0.008 \times 80 \text{ gal} \times 8.34 \text{ lb/gal}$

$$x = \frac{0.008 \times 80 \text{ gal} \times 8.34 \text{ lb/gal}}{0.11 \times 10.1 \text{ lb/gal}} = 4.9 \text{ gal}$$

92. $$\frac{(0.10 \times 32 \text{ lb}) + (0.005 \times 66 \text{ lb})}{32 \text{ lb} + 66 \text{ lb}} \times 100 = \frac{3.2 \text{ lb} + 0.33 \text{ lb}}{98 \text{ lb}} \times 100 = 3.6\%$$

93. $$\frac{(0.15\times 5\text{ gal}\times 11.2\text{ lb/gal})+(0.002\times 40\text{ gal}\times 8.34\text{ lb/gal})}{(5\text{ gal}\times 11.2\text{ lb/gal})+(40\text{ gal}\times 8.34\text{ lb/gal})}\times 100$$

$$=\frac{8.4\text{ lb}+0.67\text{ lb}}{56\text{ lb}+334\text{ lb}}\times 100=\frac{9.07\text{ lb}}{390\text{ lb}}\times 100=2.3\%$$

94. $$\frac{(0.12\times 12\text{ gal}\times 10.5\text{ lb/gal})+(0.0075\times 50\text{ gal}\times 8.40\text{ lb/gal})}{(12\text{ gal}\times 10.5\text{ lb/gal})+(40\text{ gal}\times 8.40\text{ lb/gal})}\times 100$$

$$=\frac{15.1\text{ lb}+3.2\text{ lb}}{126\text{ lb}+336\text{ lb}}\times 100=\frac{18.3\text{ lb}}{458\text{ lb}}\times 100=4.0\%$$

95. 2.3 lb ÷ 30 min = 0.08 lb/min
0.08 lb/min × 1440 min/day = 115.2 lb/day
96. 42 oz. ÷ 16 oz./lb = 2.61 lb
2.6 lb ÷ 45 min = 0.06 lb/min
0.06 lb/min × 1440 min/day = 86.4 lb/day
97. 14 oz. ÷ 16 oz./lb = 0.88 lb
(2.4 lb chemical + container) – (0.88 lb container) = 1.52 lb chemical
1.52 lb ÷ 30 min = 0.051 lb/min
0.051 lb/min × 1440 min/day = 73 lb/day
98. (2.8 lb container + chemical) – (0.6 lb container) = 2.2 lb chemical
2.2 lb ÷ 30 min = 0.073 lb/min
0.073 lb/min × 1440 min/day = 105 lb/day
99. $x\text{ mg/L}\times 1.92\text{ MGD}\times 8.34\text{ lb/gal}=42\text{ lb}$

$$x=\frac{42\text{ lb}}{1.92\text{ MGD}\times 8.34\text{ lb/gal}}=2.6\text{ mg/L}$$

100. 16,000 mg/L × 0.000070 MGD × 8.34 lb/gal = 9.3 lb/day
101. 590 mL ÷ 5 min = 118 mL/min

$$\frac{118\text{ mL/min}\times 1\text{ gal}\times 1440\text{ min/day}}{3785\text{ mL}}=44.9\text{ gpd}$$

$12{,}000\text{ mg/L}\times 0.0000449\text{ MGD}\times 8.34\text{ lb/gal}\times 1.009=4.9\text{ lb/day}$

102. 725 mL ÷ 5 min = 145 mL/min

$$\frac{145\text{ mL/min}\times 1\text{ gal}\times 1440\text{ min/day}}{3785\text{ mL}}=55\text{ gpd}$$

$12{,}000\text{ mg/L}\times 0.000055\text{ MGD}\times 8.34\text{ lb/gal}=5.5\text{ lb/day}$

103. 950 mL ÷ 5 min = 190 mL/min

$$\frac{190\text{ mL/min}\times 1\text{ gal}\times 1440\text{ min/day}}{3785\text{ mL}}=72.3\text{ gpd}$$

$14{,}000\text{ mg/L}\times 0.0000723\text{ MGD}\times 8.34\text{ lb/gal}=8.4\text{ lb/day}$

104. $1730 \text{ mL} \div 10 \text{ min} = 173 \text{ mL/min}$

$$\frac{173 \text{ mL/min} \times 1 \text{ gal} \times 1440 \text{ min/day}}{3785 \text{ mL}} = 65.8 \text{ gpd}$$

$19{,}000 \text{ mg/L} \times 0.0000658 \text{ MGD} \times 8.34 \text{ lb/gal} \times 1.009 = 11.4 \text{ lb/day}$

105. $4 \text{ in.} \div (12 \text{ in./ft}) = 0.3 \text{ ft}$

$$\frac{0.785 \times (4 \text{ ft})^2 \times 0.3 \text{ ft} \times 7.48 \text{ gal/ft}^3}{5 \text{ min}} = 5.6 \text{ gpm}$$

106. $4 \text{ in.} \div (12 \text{ in./ft}) = 0.3 \text{ ft}$

$$\frac{0.785 \times (4 \text{ ft})^2 \times 0.3 \text{ ft} \times 7.48 \text{ gal/ft}^3}{15 \text{ min}} = 1.9 \text{ gpm}$$

107. $3 \text{ in.} \div (12 \text{ in./ft}) = 0.25 \text{ ft}$

$$\frac{0.785 \times (3 \text{ ft})^2 \times 0.25 \text{ ft} \times 7.48 \text{ gal/ft}^3}{10 \text{ min}} = 1.32 \text{ gpm}$$

$1.32 \text{ gpm} \times 1440 \text{ min/day} = 1901 \text{ gpd}$

108. $2 \text{ in.} \div (12 \text{ in./ft}) = 0.17 \text{ ft}$

$$\frac{0.785 \times (3 \text{ ft})^2 \times 0.17 \text{ ft} \times 7.48 \text{ gal/ft}^3}{15 \text{ min}} = 0.6 \text{ gpm}$$

$0.6 \text{ gpm} \times 1440 \text{ min/day} = 864 \text{ gpd}$

$12{,}000 \text{ mg/L} \times 0.000864 \text{ MGD} \times 8.34 \text{ lb/gal} = 86.5 \text{ lb/day}$

109. $$\frac{0.785 \times (4 \text{ ft})^2 \times 0.17 \text{ ft} \times 7.48 \text{ gal/ft}^3}{30 \text{ min}} = 0.53 \text{ gpm}$$

$0.145 \text{ gpm} \times 1440 \text{ min/day} = 209 \text{ gpd}$

$14{,}500 \text{ mg/L} \times 0.000209 \text{ MGD} \times 8.34 \text{ lb/gal} = 25 \text{ lb/day}$

110. 535 lb × 7 days = 76.4 lb/day
111. 2200 lb × 90 lb/day = 24.4 days
112. 889 lb × 58 lb/day = 15.3 days
113. $0.785 \times (3 \text{ ft})^2 \times 3.4 \text{ ft} \times 7.48 \text{ gal/ft}^3 = 180 \text{ gal}$
180 days ÷ 88 gpd = 2 days
114. 2.8 mg/L × 1.8 MGD × 8.34 lb/gal = 42 lb/day
42 lb/day × 30 days = 1260 lb

115. $$\frac{6{,}100{,}000 \text{ gpd}}{1440 \text{ min/day} \times (60 \text{ sec/min})} = 71 \text{ gps}$$

$$\frac{3 \text{ ft} \times (4 \text{ ft})^2 \times 7.48 \text{ gal/ft}^3}{71 \text{ gps}} = 5 \text{ sec}$$

116. $50 \text{ ft} \times 20 \text{ ft} \times 9 \text{ ft} \times 7.48 \text{ gal/ft}^3 = 67{,}320 \text{ gal}$

117. $9\text{ mg/L} \times 4.35\text{ MGD} \times 8.34\text{ lb/gal} = 326\text{ lb/day}$

118. $10\text{ mg/L} \times 3.15\text{ MGD} \times 8.34\text{ lb/gal} = 500{,}000\text{ mg/L} \times x\text{ MGD} \times 8.34\text{ lb/gal}$

$$\frac{10\text{ mg/L} \times 3.15\text{ MGD} \times 8.34\text{ lb/gal}}{500{,}000\text{ mg/L} \times 8.34\text{ lb/gal}} = x\text{ MGD}$$

$$0.000063\text{ MGD} = 63.0\text{ gpd} = x$$

119. $4\text{ ft} \times 4\text{ ft} \times 2\text{ ft} \times 7.48\text{ gal/ft}^3 = 239\text{ gal}$

120. $$\frac{45\text{ gpd} \times 3785\text{ mL/gal}}{1440\text{ min/day}} = 118\text{ mL/min}$$

121. $$\frac{2{,}220{,}000}{1440\text{ min/day}} = 1542\text{ gpm}$$

$$\frac{40\text{ ft} \times 20\text{ ft} \times 9.17\text{ ft} \times 7.48\text{ gal/ft}^3}{1542\text{ gpm}} = 36\text{ min}$$

122. $8\text{ mg/L} \times 1.84\text{ MGD} \times 8.34\text{ lb/gal} = 600{,}000\text{ mg/L} \times x\text{ MGD} \times 10.2\text{ lb/gal}$

$$\frac{8\text{ mg/L} \times 1.84\text{ MGD} \times 8.34\text{ lb/gal}}{600{,}000\text{ mg/L} \times 10.2\text{ lb/gal}} = x\text{ MGD}$$

$$0.00002\text{ MGD} = 20\text{ gpd} = x$$

$$\frac{20\text{ gpd} \times 3785\text{ mL/gal}}{1440\text{ min/day}} = 52.6\text{ mL/min}$$

123. $$\frac{180\text{ gpd} \times 3785\text{ mL/gal}}{1440\text{ min/day}} = 473.1\text{ mL/min}$$

124. $6\text{ mg/L} \times 0.925\text{ MGD} \times 8.34\text{ lb/gal} = 46.3\text{ lb/day}$

125. $$\frac{2.7\text{ lb}}{x\text{ gal} \times 8.34\text{ lb/gal} + 2.7\text{ lb}} \times 100 = 1.4$$

$$\frac{270}{8.34x + 2.7\text{ lb}} = 1.4$$

$$178.6 = 8.34x + 2.7$$

$$175.9 = 8.34x$$

$$\frac{175.9}{8.34} = x$$

$$21.1\text{ gal} = x$$

126. $$\frac{(16/100 \times 25\text{ lb}) + (0.6/100 \times 100) \times 140\text{ lb}}{25\text{ lb} + 140\text{ lb}} \times 100$$

$$= \frac{4\text{ lb} + 0.84\text{ lb}}{165\text{ lb}} \times 100 = \frac{4.84\text{ lb}}{165\text{ lb}} \times 100 = 2.9\%$$

127. $\dfrac{4.0\text{ lb} \times 2}{30\text{ min} \times 2} = \dfrac{8\text{ lb}}{60\text{ min}}$, or 8 lb/hr; $8\text{ lb/hr} \times 24\text{ hr/day} = 192\text{ lb/day}$

128. $\dfrac{2.2 \text{ lb} \times 2}{30 \text{ min} \times 2} = \dfrac{4.4 \text{ lb}}{60 \text{ min}}$, or 4.4 lb/hr

$4.4 \text{ lb/hr} \times 24 \text{ hr/day} = 105.6 \text{ lb/day}$

129. 190 g = 0.0022 lb × 190 = 0.42 lb

$$\frac{0.42 \text{ lb}}{(25 \text{ gal} \times 8.34 \text{ lb/gal}) + 0.42 \text{ lb}} \times 100$$

$$= \frac{0.42 \text{ lb}}{208.5 \text{ lb} + 0.42 \text{ lb}} \times 100 = \frac{0.42 \text{ lb}}{208.9 \text{ lb}} \times 100 = 0.2\%$$

130. $\dfrac{760 \text{ mL}}{5 \text{ min}} = 152 \text{ mL/min}$

$$\frac{(152 \text{ mL/min}) \times (1440 \text{ min/day})}{3785 \text{ mL/gal}} = 58 \text{ gpd}$$

$20{,}000 \text{ mg/L} \times 0.000058 \text{ MGD} \times 8.34 \text{ lb/gal} = 9.7 \text{ lb/day}$

131. 14 mg/L × 4.2 MGD) × 8.34 lb/gal = 490 lb/day

$$\frac{490 \text{ lb/day}}{5.66 \text{ lb alum/gal solution}} = 86.6 \text{ gpd}$$

132. 0.8/10 × 210 lb = 17 lb of 10% solution
9.2/10 × 210 lb = 193 lb of water
133. 1/60 × 175 lb = 2.9 lb of 60% solution
59/60 × 75 lb = 172 lb of water
134. 3 in. ÷ 12 in./ft = 0.25 ft

$$\frac{0.785 \times (4 \text{ ft})^2 \times 0.25 \text{ ft} \times 747 \text{ gal/ft}^3}{10 \text{ min}} = 23 \text{ gal/10 min} = 2.3 \text{ gpm}$$

135. $(10/100) \times (x \text{ gal}) \times (10.2 \text{ lb/gal}) = (0.6/100) \times (80 \text{ gal}) \times (8.34 \text{ lb/gal})$

$$x = \frac{(0.006) \times (80 \text{ gal}) \times (8.34 \text{ lb/gal})}{(0.1) \times (10.2 \text{ lb/gal})} = 3.9 \text{ gal}$$

136. 710 mL ÷ 5 min = 142 mL/min

$$\frac{(142 \text{ mg/L/min}) \times (1440 \text{ min/day})}{3785 \text{ mL/gal}} = 54 \text{ gpd}$$

$9000 \text{ mg/L} \times 0.0000540 \text{ MGD} \times 8.34 \text{ lb/gal} = 4.1 \text{ lb/day}$

137. 6 mg/L × 3.7 MGD × 8.34 lb/gal = 185 lb/day
185 lb/day × 30 days = 5550 lb
138. 550 lb ÷ 80 lb/day = 6.9 days
139. 70 ft × 30 ft × 14 ft × 7.48 gal/$ft^3$ = 219,912 gal
140. 0.785 × (80 ft)$^2$ × 12 ft × 7.48 gal/$ft^3$ = 450,954 gal
141. 70 ft × 20 ft × 10 ft × 7.48 gal/$ft^3$ = 104,720 gal

142. $50{,}000 \text{ gal} = 40 \text{ ft} \times 25 \text{ ft} \times x \text{ ft} \times 7.48 \text{ gal/ft}^3$

$$\frac{50{,}000 \text{ gal}}{40 \text{ ft} \times 25 \text{ ft} \times 7.48 \text{ gal/ft}^3} = x \text{ ft}$$

$$6.7 \text{ ft} = x$$

143. $5 \text{ in.} \div 12 \text{ in./ft} = 0.42 \text{ ft}$
$0.785 \times (75 \text{ ft})^2 \times 10.42 \text{ ft} \times 7.48 \text{ gal/ft}^3 = 344{,}161 \text{ gal}$

144. $2{,}220{,}000 \div 24 \text{ hr/day} = 92{,}500 \text{ gph}$

$$\frac{70 \text{ ft} \times 25 \text{ ft} \times 10 \text{ ft} \times 7.48 \text{ gal/ft}^3}{92{,}500 \text{ gph}} = 1.4 \text{ hr}$$

145. $2{,}920{,}000 \text{ gpd} \div 24 \text{ hr/day} = 121{,}667 \text{ gph}$

$$\frac{0.785 \times (80 \text{ ft})^2 \times 12 \text{ ft} \times 7.48 \text{ gal/ft}^3}{121{,}667 \text{ gph}} = 3.7 \text{ hr}$$

146. $1{,}520{,}000 \text{ gpd} \div 24 \text{ hr/day} = 63{,}333 \text{ gph}$

$$\frac{60 \text{ ft} \times 20 \text{ ft} \times 10 \text{ ft} \times 7.48 \text{ gal/ft}^3}{63{,}333 \text{ gph}} = 1.4 \text{ hr}$$

147.

$$3 \text{ hr} = \frac{0.785 \times (60 \text{ ft})^2 \times 12 \text{ ft} \times 7.48 \text{ gal/ft}^3}{x \text{ gph}}$$

$$x = \frac{0.785 \times (60 \text{ ft})^2 \times 12 \text{ ft} \times 7.48 \text{ gal/ft}^3}{3 \text{ hr}} = 84{,}554 \text{ gph}$$

$84{,}554 \text{ gph} \times 24 \text{ hr/day} = 2{,}029{,}296 \text{ gpd} = 2.0 \text{ MGD}$

148. $1{,}740{,}000 \text{ gpd} \div 24 \text{ hr/day} = 72{,}500 \text{ gph}$

$$\frac{70 \text{ ft} \times 25 \text{ ft} \times 12 \text{ ft} \times 7.48 \text{ gal/ft}^3}{72{,}500 \text{ gph}} = 2.2 \text{ hr}$$

149. $$\frac{510 \text{ gpm}}{60 \text{ ft} \times 25 \text{ ft}} = 0.34 \text{ gpm/ft}^2$$

150. $$\frac{1610 \text{ gpm}}{0.785 \times 70 \text{ ft} \times 70 \text{ ft}} = 0.42 \text{ gpm/ft}^2$$

151. $540{,}000 \text{ gpd} \div 1440 \text{ min/day} = 375 \text{ gpm}$

$$\frac{375 \text{ gpm}}{50 \text{ ft} \times 20 \text{ ft}} = 0.38 \text{ gpm/ft}^2$$

152.

$$0.5 \text{ gpm/ft}^2 = \frac{x \text{ gpm}}{80 \text{ ft} \times 25 \text{ ft}}$$

$$0.5 \text{ gpm/ft}^2 \times 80 \text{ ft} \times 25 \text{ ft} = x \text{ gpm}$$

$$1000 \text{ gpm} = x$$

$1000 \text{ gpm} \times 1440 \text{ min/day} = 1{,}440{,}000 \text{ gpd}$

153. 1,820,000 gpd ÷ 1440 min/day = 1264 gpm

$$\frac{1264 \text{ gpm}}{0.785 \times 60 \text{ ft} \times 60 \text{ ft}} = 0.45 \text{ gpm/ft}^2$$

154.
$$\frac{1,550,000 \text{ gpd}}{1440 \text{ min/day} \times 7.48 \text{ gal/ft}^3} = 144 \text{ cfm}$$

$$144 \text{ cfm} = 25 \text{ ft} \times 12 \text{ ft} \times x \text{ fpm}$$

$$\frac{144 \text{ cfm}}{25 \text{ ft} \times 12 \text{ ft}} = x \text{ fpm}$$

$$0.5 \text{ fpm} = x$$

155.
$$\frac{1,800,000 \text{ gpd}}{1440 \text{ min/day} \times 7.48 \text{ gal/ft}^3} = 167 \text{ cfm}$$

$$167 \text{ cfm} = 30 \text{ ft} \times 12 \text{ ft} \times x \text{ fpm}$$

$$\frac{167 \text{ cfm}}{30 \text{ ft} \times 12 \text{ ft}} = x \text{ fpm}$$

$$0.5 \text{ fpm} = x$$

156.
$$\frac{2,450,000 \text{ gpd}}{1440 \text{ min/day} \times 7.48 \text{ gal/ft}^3} = 227 \text{ cfm}$$

$$227 \text{ cfm} = 30 \text{ ft} \times 14 \text{ ft} \times x \text{ fpm}$$

$$\frac{227 \text{ cfm}}{30 \text{ ft} \times 14 \text{ ft}} = x \text{ fpm}$$

$$0.5 \text{ fpm} = x$$

157.
$$\frac{2,880,000 \text{ gpd}}{1440 \text{ min/day} \times 7.48 \text{ gal/ft}^3} = 267 \text{ cfm}$$

$$267 \text{ cfm} = 40 \text{ ft} \times 10 \text{ ft} \times x \text{ fpm}$$

$$\frac{267 \text{ cfm}}{40 \text{ ft} \times 12 \text{ ft}} = x \text{ fpm}$$

$$0.56 \text{ fpm} = x$$

158.
$$\frac{910,000 \text{ gpd}}{1440 \text{ min/day} \times 7.48 \text{ gal/ft}^3} = 84.5 \text{ cfm}$$

$$84.5 \text{ cfm} = 25 \text{ ft} \times 10 \text{ ft} \times x \text{ fpm}$$

$$\frac{84.5 \text{ cfm}}{25 \text{ ft} \times 10 \text{ ft}} = x \text{ fpm}$$

$$0.4 \text{ fpm} = x$$

159. 2,520,000 gpd ÷ 1440 min/day = 1750 gpm
1750 gpm ÷ (3.14 × 70 ft) = 7.9 gpm/ft

160. 1,890,000 gpd ÷ 1440 min/day = 1313 gpm
 1313 gpm ÷ 170 ft = 7.7 gpm/ft
161. 1,334,000 gpd ÷ 1440 min/day = 926 gpm
 926 gpm ÷ 120 ft = 7.7 gpm/ft
162. 3,700,000 gpd ÷ 1440 min/day = 2569 gpm
 2569 gpm ÷ (3.14 × 70 ft) = 11.7 gpm/ft
163. 1,900,000 gpd ÷ 1440 min/day = 1319 gpm
 1319 gpm ÷ 160 ft = 8.2 gpm/ft
164. (22 mL ÷ 100 mL) × 100 = 22%
165. (25 mL ÷ 100 mL) × 100 = 25%
166. (15 mL ÷ 100 mL) × 100 = 15%
167. (16 mL ÷ 100 mL) × 100 = 16%

168.
$$\frac{0.45 \text{ mg/L}}{1 \text{ mg/L}} = \frac{x \text{ mg/L}}{52 \text{ mg/L}}$$

$$0.45 \text{ mg/L} \times 52 \text{ mg/L} = 1 \text{ mg/L} \times x \text{ mg/L}$$

$$23.4 \text{ mg/L} = x$$

23.4 mg/L + 40 mg/L = 63.4 mg/L

169.
$$\frac{0.45 \text{ mg/L}}{1 \text{ mg/L}} = \frac{x \text{ mg/L}}{60 \text{ mg/L}}$$

$$0.45 \text{ mg/L} \times 60 \text{ mg/L} = 1 \text{ mg/L} \times x \text{ mg/L}$$

$$27 \text{ mg/L} = x$$

27 mg/L + 30 mg/L = 57 mg/L

170. 40 mg/L – 26 mg/L = 14 mg/L alkalinity
171. 40 mg/L – 28 mg/L = 12 mg/L alkalinity

172.
$$\frac{0.45 \text{ mg/L}}{0.45 \text{ mg/L}} = \frac{15 \text{ mg/L}}{x \text{ mg/L}}$$

$$0.45 \text{ mg/L} \times x \text{ mg/L} = 0.45 \text{ mg/L} \times 15 \text{ mg/L}$$

$$x \text{ mg/L} = \frac{0.45 \text{ mg/L} \times 15 \text{ mg/L}}{0.45 \text{ mg/L}}$$

$$x = 15 \text{ mg/L lime}$$

173.
$$\frac{0.45 \text{ mg/L}}{0.35 \text{ mg/L}} = \frac{20 \text{ mg/L}}{x \text{ mg/L}}$$

$$0.45 \text{ mg/L} \times x \text{ mg/L} = 0.35 \text{ mg/L} \times 20 \text{ mg/L}$$

$$x \text{ mg/L} = \frac{0.35 \text{ mg/L} \times 20 \text{ mg/L}}{0.45 \text{ mg/L}}$$

$$x = 15.6 \text{ mg/L lime}$$

174. $$\frac{0.45\text{ mg/L}}{1\text{ mg/L}} = \frac{x\text{ mg/L}}{55\text{ mg/L}}$$

$$0.45\text{ mg/L} \times 55\text{ mg/L} = 1\text{ mg/L} \times x\text{ mg/L}$$

$$24.8\text{ mg/L alkalinity} = x$$

24.8 mg/L + 30 mg/L = 54.8 mg/L total alkalinity required
54.8 mg/L – 35 mg/L = 19.8 mg/L alkalinity to be added to the water

$$\frac{0.45\text{ mg/L}}{0.35\text{ mg/L}} = \frac{19.8\text{ mg/L}}{x\text{ mg/L}}$$

$$0.45\text{ mg/L} \times x\text{ mg/L} = 0.35\text{ mg/L} \times 19.8\text{ mg/L}$$

$$x = \frac{0.35\text{ mg/L} \times 19.8\text{ mg/L}}{0.45\text{ mg/L}}$$

$$= 15.4\text{ mg/L lime}$$

175. 13.8 mg/L × 2.7 MGD × 8.34 lb/gal = 311 lb/day lime
176. 12.3 mg/L × 2.24 MGD × 8.34 lb/gal = 230 lb/day lime
177. 16.1 mg/L × 0.99 MGD × 8.34 lb/gal = 133 lb/day lime
178. 15 mg/L × 2.2 MGD × 8.34 lb/gal = 275 lb/day lime

179. $$\frac{205\text{ lb/day} \times 453.6\text{ g/lb}}{1440\text{ min/day}} = 64.6\text{ g/min lime}$$

180. $$\frac{110\text{ lb/day} \times 453.6\text{ g/lb}}{1440\text{ min/day}} = 34.7\text{ g/min lime}$$

181. 12 mg/L × 0.90 MGD × 8.34 lb/gal = 90 lb/day lime

$$\frac{90\text{ lb/day} \times 453.6\text{ g/lb}}{1440\text{ min/day}} = 28.4\text{ g/min lime}$$

182. 14 mg/L × 2.66 MGD × 8.34 lb/gal = 310.1 lb/day lime

$$\frac{310.1\text{ lb/day} \times 453.6\text{ g/lb}}{1440\text{ min/day}} = 97.7\text{ g/min lime}$$

183. 1,550,000 gpd ÷ 24 hr/day = 64,583 gph

$$\frac{66\text{ ft} \times 30\text{ ft} \times 12\text{ ft} \times 7.48\text{ gal/ft}^3}{64{,}583\text{ gph}} = 2.8\text{ hr}$$

184. 70 ft × 30 ft × 7.48 gal/ft$^3$ = 219,912 gal

185. $$\frac{1{,}620{,}000\text{ gpd}}{1440\text{ min/day} \times 7.48\text{ gal/ft}^3} = 150\text{ cfm}$$

$$150\text{ cfm} = 25\text{ ft} \times 12\text{ ft} \times x\text{ fpm}$$

$$\frac{150\text{ cfm}}{25\text{ ft} \times 12\text{ ft}} = x\text{ fpm}$$

$$0.5\text{ fpm} = x$$

186. 635,000 gpd ÷ 1440 min/day = 441 gpm
441 gpm ÷ (40 ft × 25 ft) = 0.44 gpm

187. 0.785 × 70 ft × 70 ft × 14 ft × 7.48 gal/ft³ = 402,805 gal

188. 2,220,000 gpd ÷ 1440 min/day = 1542 gpm
1542 gpm ÷ 180 ft = 8.6 gpm/ft

189. 2,560,000 gpd ÷ 24 hr/day = 106,667 gph

$$\frac{0.785 \times (60\text{ ft})^2 \times 10\text{ ft} \times 7.48\text{ gal/ft}^3}{106{,}667\text{ gph}} = 1.98\text{ hr}$$

190. $$\frac{1{,}750{,}000\text{ gpd}}{1440\text{ min/day} \times 7.48\text{ gal/ft}^3} = 162\text{ cfm}$$

$$162\text{ cfm} = 30\text{ ft} \times 12\text{ ft} \times x\text{ fpm}$$

$$\frac{162\text{ cfm}}{30\text{ ft} \times 12\text{ ft}} = x\text{ fpm}$$

$$0.45\text{ fpm} = x$$

191. $$\frac{1700\text{ gpm}}{0.785 \times 70\text{ ft} \times 70\text{ ft}} = 0.4\text{ gpm/ft}^2$$

192. 3,150,000 gpd ÷ 1440 min/day = 2188 gpm

$$\frac{2188\text{ gpm}}{3.14\text{ ft} \times 70\text{ ft}} = 9.9\text{ gpm}$$

193. $$2\text{ hr} = \frac{0.785 \times (60\text{ ft})^2 \times 12\text{ ft} \times 7.48\text{ gal/ft}^3}{x\text{ gph}}$$

$$x = \frac{0.785 \times (60\text{ ft})^2 \times 12\text{ ft} \times 7.48\text{ gal/ft}^3}{2\text{ hr}} = 126{,}831\text{ gph}$$

126,831 gph × 24 hr/day = 3,043,944 gpd (3.04 MGD)

194. $$\frac{3{,}250{,}000\text{ gpd}}{1440\text{ min/day} \times 7.48\text{ gal/ft}^3} = 302\text{ cfm}$$

$$302\text{ cfm} = 30\text{ ft} \times 14\text{ ft} \times x\text{ fpm}$$

$$\frac{302\text{ cfm}}{30\text{ ft} \times 14\text{ ft}} = x\text{ fpm}$$

$$0.7\text{ fpm} = x$$

195. (26 mL ÷ 100 mL) × 100 = 26%

196. $$\frac{1\text{ mg/L}}{0.45\text{ mg/L}} = \frac{50\text{ mg/L}}{x\text{ mg/L}}$$

$$1\text{ mg/L} \times x\text{ mg/L} = 0.45\text{ mg/L} \times 50\text{ mg/L}$$

$$x = 22.5\text{ mg/L}$$

Total alkalinity required = 22.5 mg/L + 30 mg/L = 52.5 mg/L

197. $$0.7\text{ gpm/ft}^2 = \frac{x\text{ gpm}}{80\text{ ft} \times 30\text{ ft}}$$

$$0.7\text{ gpm/ft}^2 \times 80\text{ ft} \times 30\text{ ft} = x\text{ gpm}$$

$$1680\text{ gpm} = x$$

1680 gpm × 1440 min/day = 2,419,200 gpd

198. 14.5 mg/L × 2.41 MGD × 8.34 lb/gal = 291 lb/day

199. (21 mL ÷ 100 mL) × 100 = 21%

200. 3,240,000 gpd ÷ 1440 min/day = 2250 gpm
2250 gpm ÷ (3.14 × 80 ft) = 90 gpm/ft

201. 50 mg/L – 30 mg/L = 10 mg/L alkalinity to be added to the water

202. $$\frac{0.45\text{ mg/L}}{1\text{ mg/L}} = \frac{x\text{ mg/L}}{50\text{ mg/L}}$$

$$0.45\text{ mg/L} \times 50\text{ mg/L} = 1\text{ mg/L} \times x\text{ mg/L}$$

$$22.5\text{ mg/L alkalinity} = x$$

22.5 mg/L + 30 mg/L = 52.5 mg/L alkalinity required
52.5 mg/L – 33 mg/L = 19.5 mg/L alkalinity to be added

$$\frac{0.45\text{ mg/L}}{0.35\text{ mg/L}} = \frac{19.5\text{ mg/L}}{x\text{ mg/L}}$$

$$0.45\text{ mg/L} \times x\text{ mg/L} = 0.35\text{ mg/L} \times 19.5\text{ mg/L}$$

$$x = \frac{0.35\text{ mg/L} \times 19.5\text{ mg/L}}{0.45\text{ mg/L}} = 15.2\text{ mg/L lime}$$

203. $$\frac{192\text{ lb/day} \times 453.6\text{ g/lb}}{1440\text{ min/day}} = 60.5\text{ g/min}$$

204. 16 mg/L × 1.5 MGD × 8.34 lb/gal = 200 lb/day

205. $$\text{Lime (g/min)} = \frac{\text{Lime (lb/day)} \times 453.6\text{ g/lb}}{1440\text{ min/day}}$$

$$= \frac{14\text{ mg/L} \times 2.88\text{ MGD} \times 8.34\text{ lb/gal} \times 453.6\text{ g/lb}}{1440\text{ min/day}}$$

$$= 106\text{ g/min}$$

206. $$\frac{14{,}200{,}000\text{ gal}}{80\text{ hr} \times 60\text{ min/lb}} = 2958\text{ gpm}$$

207. 2,970,000 gpd ÷ 1440 min/day = 2063 gpm

208. $$3200\text{ gpm} = \frac{16{,}000{,}000\text{ gal}}{x\text{ hr} \times 60\text{ min/hr}}$$

$$x = \frac{16{,}000{,}000\text{ gal}}{3200\text{ gpm} \times 60\text{ min/hr}} = 83\text{ hr}$$

209. 45 ft × 22 ft × (1ft/5 min) × 7.48 gal/ft$^3$ = 1481 gpm

210. 14 in. ÷ 12 in./ft = 1.17 ft
 40 ft × 30 ft × (1.17 ft/5 min) × 7.48 gal/ft³ = 2100 gpm
211. 18 in. ÷ 12 in./ft = 1.5 ft
 35 ft × 18 ft × (1.5 ft/6 min) × 7.48 gal/ft³ = 1178 gpm
212. $\frac{1760 \text{ gpm}}{20 \text{ ft} \times 18 \text{ ft}} = 4.9 \text{ gpm/ft}^2$
213. 2,150,000 gal ÷ 1440 min/day = 1493 gpm
 $\frac{1493 \text{ gpm}}{32 \text{ ft} \times 18 \text{ ft}} = 2.6 \text{ gpm/ft}^2$
214. $\frac{18{,}100{,}000 \text{ gal}}{71.6 \text{ hr} \times 60 \text{ min/hr}} = 4213 \text{ gpm}$
 $\frac{4213 \text{ gpm}}{38 \text{ ft} \times 24 \text{ ft}} = 4.6 \text{ gpm/ft}^2$
215. $\frac{14{,}200{,}000 \text{ gal}}{71.4 \text{ hr} \times 60 \text{ min/hr}} = 3315 \text{ gpm}$
 $\frac{3315 \text{ gpm}}{33 \text{ ft} \times 24 \text{ ft}} = 4.2 \text{ gpm/ft}^2$
216. 3,550,000 gpd ÷ 1440 min/day = 2465 gpm
 $\frac{2465 \text{ gpm}}{88 \text{ ft} \times 22 \text{ ft}} = 2.9 \text{ gpm/ft}^2$
217. 22 in. ÷ 12 in./ft = 1.83 ft
 $\frac{1873 \text{ gpm}}{38 \text{ ft} \times 18 \text{ ft}} = 2.7 \text{ gpm/ft}^2$
218. 21 in. ÷ 12 in./ft = 1.8 ft
 33 ft × 24 ft × (1.8 ft/6 min) × 7.48 gal/ft³ = 1777 gpm
 $\frac{1777 \text{ gpm}}{33 \text{ ft} \times 24 \text{ ft}} = 2.2 \text{ gpm/ft}^2$
219. $\frac{2{,}870{,}000 \text{ gal}}{20 \text{ ft} \times 18 \text{ ft}} = 7972 \text{ gal/ft}^2$
220. $\frac{4{,}180{,}000 \text{ gal}}{32 \text{ ft} \times 20 \text{ ft}} = 6533 \text{ gal/ft}^2$
221. $\frac{2{,}980{,}000 \text{ gal}}{24 \text{ hr} \times 18 \text{ ft}} = 6898 \text{ gal/ft}^2$
222. 3.4 gpm/ft² × 3330 min = 11,322 gal/ft²
223. 2.6 gpm/ft² × 60.5 hr × 60 min/hr = 9438 gal/ft²
224. 3510 gpm ÷ 380 ft² = 9.2 gpm/ft²
225. $\frac{3580 \text{ gpm}}{18 \text{ ft} \times 14 \text{ ft}} = 14.2 \text{ gpm/ft}^2$

226. $(16 \text{ gpm/ft}^2 \times 12 \text{ in./ft}) \div 7.48 \text{ gal/ft}^2 = 25.7 \text{ in./min}$

227. $$\frac{3650 \text{ gpm}}{30 \text{ ft} \times 18 \text{ ft}} = 6.8 \text{ gpm/ft}^2$$

228. $$\frac{3080 \text{ gpm}}{18 \text{ ft} \times 14 \text{ ft}} = 12.2 \text{ gpm/ft}^2$$

$12.2 \text{ gpm/ft}^2 \times 1.6 = 19.5 \text{ in./min rise}$

229. $6650 \text{ gpm} \times 6 \text{ min} = 39{,}900 \text{ gal}$
230. $9100 \text{ gpm} \times 7 \text{ min} = 63{,}700 \text{ gal}$
231. $4670 \text{ gpm} \times 5 \text{ min} = 23{,}350 \text{ gal}$
232. $6750 \text{ gpm} \times 6 \text{ min} = 40{,}500 \text{ gal}$

233. $59{,}200 \text{ gal} = 0.785 \times (40 \text{ ft})^2 \times x \text{ ft} \times 7.48 \text{ gal/ft}^3$

$$\frac{59{,}200 \text{ gal}}{0.785 \times (40 \text{ ft})^2 \times 7.48 \text{ gal/ft}^3} = x \text{ ft}$$

$$6.3 \text{ ft} = x$$

234. $62{,}200 \text{ gal} = 0.785 \times (52 \text{ ft})^2 \times x \text{ ft} \times 7.48 \text{ gal/ft}^3$

$$\frac{62{,}200 \text{ gal}}{0.785 \times (52 \text{ ft})^2 \times 7.48 \text{ gal/ft}^3} = x \text{ ft}$$

$$3.9 \text{ ft} = x$$

235. $42{,}300 \text{ gal} = 0.785 \times (42 \text{ ft})^2 \times x \text{ ft} \times 7.48 \text{ gal/ft}^3$

$$\frac{42{,}300 \text{ gal}}{0.785 \times (42 \text{ ft})^2 \times 7.48 \text{ gal/ft}^3} = x \text{ ft}$$

$$4.1 \text{ ft} = x$$

236. $7150 \text{ gpm} \times 7 \text{ min} = 50{,}050 \text{ gal}$

$50{,}050 \text{ gal} = 0.785 \times (40 \text{ ft})^2 \times x \text{ ft} \times 7.48 \text{ gal/ft}^3$

$$\frac{50{,}050 \text{ gal}}{0.785 \times (40 \text{ ft})^2 \times 7.48 \text{ gal/ft}^3} = x \text{ ft}$$

$$5.3 \text{ ft} = x$$

237. $8860 \text{ gpm} \times 6 \text{ min} = 53{,}160 \text{ gal}$

$53{,}160 \text{ gal} = 0.785 \times (40 \text{ ft})^2 \times x \text{ ft} \times 7.48 \text{ gal/ft}^3$

$$\frac{53{,}160 \text{ gal}}{0.785 \times (40 \text{ ft})^2 \times 7.48 \text{ gal/ft}^3} = x \text{ ft}$$

$$5.7 \text{ ft} = x$$

238. $19 \text{ gpm/ft}^2 \times 42 \text{ ft} \times 22 \text{ ft} = 17{,}556 \text{ gpm}$
239. $20 \text{ gpm/ft}^2 \times 36 \text{ ft} \times 26 \text{ ft} = 18{,}720 \text{ gpm}$
240. $16 \text{ gpm/ft}^2 \times 22 \text{ ft} \times 22 \text{ ft} = 7744 \text{ gpm}$
241. $24 \text{ gpm/ft}^2 \times 26 \text{ ft} \times 22 \text{ ft} = 13{,}728 \text{ gpm}$

242. (74,200 gal ÷ 17,100,000 gal) × 100 = 0.43%
243. (37,200 gal ÷ 6,100,000 gal) × 100 = 0.61%
244. (59,400 gal ÷ 13,100,000 gal) × 100 = 0.45%
245. (52,350 gal ÷ 11,110,000 gal) × 100 = 0.47%
246. Mud ball volume = 635 mL – 600 mL = 35 mL
(35 mL ÷ 3625 mL) × 100 = 0.97%
247. Mud ball volume = 535 mL – 510 mL = 25 mL
Total sample volume = 5 × 705 mL = 3525 mL
(25 mL ÷ 3525 mL) × 100 = 0.7%
248. Mud ball volume = 595 mL – 520 mL = 75 mL
Total sample volume = 5 × 705 mL = 3525 mL
(75 mL ÷ 3525 mL) × 100 = 2.2%
249. Mud ball volume = 562 mL – 520 mL = 42 mL
Total sample volume = 5 × 705 mL = 3525 mL
(42 mL ÷ 3525 mL) × 100 = 1.2%

250. $$\frac{11{,}400{,}000 \text{ gal}}{80 \text{ hr} \times 60 \text{ min/hr}} = 2375 \text{ gpm}$$

251. 3,560,000 gpd ÷ 1440 min/day = 2472 gpm
2472 gpm ÷ (40 ft × 25 ft) = 2.5 gpm/ft$^2$

252. $$\frac{2{,}880{,}000 \text{ gal}}{25 \text{ ft} \times 25 \text{ ft}} = 4608 \text{ gal/ft}^2$$

253. $$2900 \text{ gpm} = \frac{14{,}800{,}000 \text{ gal}}{x \text{ hr} \times 60 \text{ min/hr}}$$

$$x = \frac{14{,}800{,}000 \text{ gal}}{2900 \text{ gpm} \times 60 \text{ min/hr}} = 85.1 \text{ hr}$$

254. 14 in. ÷ 12 in./ft = 1.17 ft
38 ft × 26 ft × 1.17 ft/5 min = 231 cfm
231 cfm × 7.48 gal/ft$^3$ = 1728 gpm

255. $$\frac{3{,}450{,}000 \text{ gal}}{30 \text{ ft} \times 25 \text{ ft}} = 4600 \text{ gal/ft}^2$$

256. $$\frac{13{,}500{,}000 \text{ gal}}{73.8 \text{ hr} \times 60 \text{ min/hr}} = 3049 \text{ gpm}$$

$$\frac{3049 \text{ gpm}}{30 \text{ ft} \times 20 \text{ ft}} = 5.1 \text{ gpm/ft}^2$$

257. 3220 gpm ÷ 360 ft$^2$ = 8.9 gpm/ft$^2$
258. 6350 gpm × 6 min = 38,100 gal
259. 14 in. ÷ 12 in./ft = 1.2 ft
30 ft × 22 ft × (1.2 ft/5 min) × 7.48 gal/ft$^3$ = 1185 gpm
1185 gpm ÷ (30 ft × 22 ft) = 1.8 gpm/ft$^2$

260. $0.785 \times (45\text{ ft})^2 \times x\text{ ft} \times 7.48\text{ gal/ft}^3 = 53{,}200\text{ gal}$

$$x = \frac{53{,}200\text{ gal}}{0.785 \times (45\text{ ft})^2 \times 7.48\text{ gal/ft}^3} = 4.5\text{ ft}$$

261. $3.3\text{ gpm/ft}^2 \times 3620\text{ min} = 11{,}946\text{ gal/ft}^2$

262. 20 in. ÷ 12 in./ft = 1.7 ft
$40\text{ ft} \times 25\text{ ft} \times (1.7\text{ ft/5 min}) \times 7.48\text{ gal/ft}^3 = 2543\text{ gpm}$

$$\frac{2543\text{ gpm}}{40\text{ ft} \times 25\text{ ft}} = 2.5\text{ gpm/ft}^2$$

263. $$\frac{3800\text{ gpm}}{35\text{ ft} \times 25\text{ ft}} = 4.3\text{ gpm/ft}^2$$

264. 4500 gpm × 7 min = 31,500 gal

265. $16\text{ gpm/ft}^2 \times 30\text{ ft} \times 30\text{ ft} = 14{,}400\text{ gpm}$

266. $$\frac{2800\text{ gpm}}{25\text{ ft} \times 20\text{ ft}} = 5.6\text{ gpm/ft}^2$$

$$\frac{5.6\text{ gpm/ft}^2 \times 12\text{ in./ft}}{7.48\text{ gal/ft}^3} = 8.9\text{ in./min}$$

267. 18 in. ÷ 12 in./ft = 1.5 ft
$30\text{ ft} \times 25\text{ ft} \times 1.5/6\text{ min} \times 7.48\text{ gal/ft}^3 = 1403\text{ gpm}$

$$\frac{1403\text{ gpm}}{30\text{ ft} \times 25\text{ ft}} = 1.9\text{ gpm/ft}^2$$

268. $18\text{ gpm/ft}^2 \times 45\text{ ft} \times 25\text{ ft} = 20{,}250\text{ gpm}$

269. (71,350 gal ÷ 18,200,000 gal) × 100 = 0.39%

270. $0.785 \times (35\text{ ft})^2 \times x\text{ ft} \times 7.48\text{ gal/ft}^3 = 86{,}400\text{ gal}$

$$x = \frac{86{,}400\text{ gal}}{0.785 \times (35\text{ ft})^2 \times 7.48\text{ gal/ft}^3} = 12\text{ ft}$$

271. 527 mL – 500 mL = 27 mL
(27 mL ÷ 3480 mL) × 100 = 0.8%

272. (51,200 gal ÷ 13,800,000 gal) × 100 = 0.37%

273. 571 – 500 mL = 71 mL

$$\frac{71\text{ mL}}{5 \times 695\text{ mL}} \times 100 = 2\%$$

274. (a) $$\frac{3{,}700{,}000\text{ gal}}{36\text{ hr/500 ft}^2} \times \frac{1\text{ hr}}{60\text{ min}} = 3.4\text{ gpm/ft}^2$$

(b) $12\text{ gpm/ft}^2 \times 15\text{ min} \times 500\text{ ft}^2 = 90{,}000\text{ gal}$

(c) (90,000 gal ÷ 3,700,000 gal) × 100 = 2.4%

(d) 70 ft – 25 ft = 45 ft

(e) $3{,}700{,}000 \div 500\text{ ft}^2) \times 100 = 7400\text{ gal/ft}^2$

275. 1.8 mg/L × 3.5 MGD × 8.34 lb/gal = 52.5 lb/day
276. 2.5 mg/L × 1.34 MGD × 8.34 lb/gal = 28 lb/day
277. 0.785 × (1 ft)$^2$ × 1200 ft × 7.48 gal/ft$^3$ = 7046 gal
52 mg/L × 0.007046 MG × 8.34 lb/gal = 3.1 lb
278. $x \text{ mg/L} \times 3.35 \text{ MGD} \times 8.34 \text{ lb/gal} = 43 \text{ lb}$

$$x = \frac{43 \text{ lb}}{3.35 \text{ MGD} \times 8.34 \text{ lb/gal}} = 1.5 \text{ mg/L}$$

279. 19,222,420 gpd – 18,815,108 gpd = 407,312 gpd, or 0.407 MGD
16 lb ÷ (0.407 MGD × 8.34 lb/gal) = 4.7 mg/L
280. 1.6 mg/L + 0.5 mg/L = 2.1 mg/L
281. 2.9 mg/L = $x$ mg/L + 0.7 mg/L
2.9 mg/L – 0.7 mg/L = $x$ mg/L
2.2 mg/L = $x$
282. 2.6 mg/L + 0.8 mg/L = 3.4 mg/L
(3.4 mg/L × 3.85 MGD × 8.34 lb/gal = 109 lb/day
283. $x \text{ mg/L} \times 1.10 \text{ MGD} \times 8.34 \text{ lb/gal} = 6 \text{ lb/day}$

$$x = \frac{6 \text{ lb/day}}{1.10 \text{ MGD} \times 8.34 \text{ lb/gal}} = 0.65 \text{ mg/L}$$

0.8 mg/L – 0.65 mg/L = 0.15 mg/L
Expected increase in residual was 0.65 mg/L whereas the actual increase in residual was only 0.15 mg/L. From this analysis it appears that the water is not being chlorinated beyond the breakpoint.
284. $x \text{ mg/L} \times 2.10 \text{ MGD} \times 8.34 \text{ lb/gal} = 5 \text{ lb/day}$

$$x = \frac{5 \text{ lb/day}}{2.10 \text{ MGD} \times 8.34 \text{ lb/gal}} = 0.29 \text{ mg/L}$$

0.5 mg/L – 0.4 mg/L = 0.1 mg/L
The expected chlorine residual increase (0.29 mg/L) is consistent with the actual increase in chlorine residual (0.1 mg/L); thus, it appears as though the water is not being chlorinated beyond the breakpoint.
285. 48 lb/day ÷ 0.65 = 73.8 lb/day hypochlorite
286. 42 lb/day ÷ 0.65 = 64.6 lb/day hypochlorite
287. 2.7 mg/L × 0.928 MGD × 8.34 lb/gal = 21 lb/day chlorine
21 lb/day ÷ 0.65 = 32.3 lb/day hypochlorite
288. $$54 \text{ lb/day} = \frac{x \text{ lb/day}}{0.65}$$

$$54 \text{ lb/day} \times 0.65 = x \text{ lb/day}$$

$$35.1 \text{ lb/day} = x$$

$$x \text{ mg/L} \times 1.512 \text{ MGD} \times 8.34 \text{ lb/gal} = 35.1 \text{ lb/day}$$

$$x = \frac{35.1 \text{ lb/day}}{1.512 \text{ MGD} \times 8.34 \text{ lb/gal}} = 2.8 \text{ mg/L}$$

289. $49\ \text{lb/day} = \dfrac{x\ \text{lb/day}}{0.65}$

$$49\ \text{lb/day} \times 0.65 = x$$

$$31.9\ \text{lb/day} = x$$

$$x\ \text{mg/L} \times 3.210\ \text{MGD} \times 8.34\ \text{lb/gal} = 31.9\ \text{lb/day}$$

$$x = \frac{31.9\ \text{lb/day}}{3.210\ \text{MGD} \times 8.34\ \text{lb/gal}} = 1.2\ \text{mg/L}$$

290. 36 lb/day ÷ 8.34 lb/gal = 4.3 gpd

291. $2.2\ \text{mg/L} \times 0.245\ \text{MGD} \times 8.34\ \text{lb/gal} = 30{,}000\ \text{mg/L} \times x\ \text{MGD} \times 8.34\ \text{lb/gal}$

$$\frac{2.2\ \text{mg/L} \times 0.245\ \text{MGD} \times 8.34\ \text{lb/gal}}{30{,}000\ \text{mg/L} \times 8.34\ \text{lb/gal}} = x$$

$$0.0000179\ \text{MGD, or } 17.9\text{gpd} = x$$

292. $\dfrac{2{,}330{,}000\ \text{gal}}{7\ \text{days}} = 332{,}857\ \text{gpd}$

$$\frac{0.785 \times (3\ \text{ft})^2 \times 2.83\ \text{ft} \times 7.48\ \text{gal/ft}^3}{7\ \text{days}} = 21.4\ \text{gpd}$$

$$x\ \text{mg/L} \times 0.332\ \text{MGD} \times 8.34\ \text{lb/gal} = 30{,}000\ \text{mg/L} \times 0.0000214\ \text{MGD} \times 8.34\ \text{lb/gal}$$

$$x = \frac{30{,}000\ \text{mg/L} \times 0.0000214\ \text{MGD} \times 8.34\ \text{lb/gal}}{0.332\ \text{MGD} \times 8.34\ \text{lb/gal}}$$

$$x = 1.9\ \text{mg/L}$$

293. 400 gpm × 1440 min/day = 576,000 gpd = 0.576 MGD

$$1.8\ \text{mg/L} \times 0.576\ \text{MGD} \times 8.34\ \text{lb/gal} = 40{,}000\ \text{mg/L} \times x\ \text{MGD} \times 8.34\ \text{lb/gal}$$

$$\frac{1.8\ \text{mg/L} \times 0.576\ \text{MGD} \times 8.34\ \text{lb/gal}}{40{,}000\ \text{mg/L} \times 8.34\ \text{lb/gal}} = x$$

$$0.0000259\ \text{MGD, or } 25.9\ \text{gpd} = x$$

294. $2.9\ \text{mg/L} \times 0.955\ \text{MGD} \times 8.34\ \text{lb/gal} = 30{,}000\ \text{mg/L} \times x\ \text{MGD} \times 8.34\ \text{lb/gal}$

$$\frac{2.9\ \text{mg/L} \times 0.955\ \text{MGD} \times 8.34\ \text{lb/gal}}{30{,}000\ \text{mg/L} \times 8.34\ \text{lb/gal}} = x$$

$$0.0000923\ \text{MGD, or } 92.3\ \text{gpd} = x$$

295. $\dfrac{22\ \text{lb} \times 0.65}{(60\ \text{gal} \times 8.34\ \text{lb/gal}) + (22\ \text{lb} \times 0.65)} \times 100$

$$= \frac{14.3\ \text{lb}}{500\ \text{lb} + 14.3\ \text{lb}} \times 100 = \frac{14.3\ \text{lb}}{514.3\ \text{lb}} \times 100 = 2.7\%$$

296. 320 g $\times$ 0.0022 lb/g = 0.70 lb hypochlorite

$$\frac{0.70\text{ lb}\times 0.65}{(7\text{ gal}\times 8.34\text{ lb/gal})+(0.70\text{ lb}\times 0.65)}\times 100$$

$$=\frac{0.46\text{ lb}}{58.4\text{ lb}+0.46\text{ lb}}\times 100=\frac{0.46\text{ lb}}{58.9\text{ lb}}\times 100=0.79\%$$

297.

$$\frac{x\text{ lb}\times 0.65}{(65\text{ gal}\times 8.34\text{ lb/gal})+(x\text{ lb}\times 0.65)}\times 100=\frac{0.65x\times 100}{542.1\text{ lb/gal}+0.65x}$$

$$\frac{65x}{542.1\text{ lb/gal}+0.65x}=21.7x$$

$$542.8=21.7x$$

$$\frac{542.8}{21.7}=x$$

$$25\text{ lb}=x$$

298. $x\text{ gal}\times 8.34\text{ lb/gal}\times(10/100)=35\text{ gal}\times 8.34\text{ lb/gal}\times(2/100)$

$$x\text{ gal}=\frac{35\text{ gal}\times 8.34\text{ lb/gal}\times 0.02}{8.34\text{ lb/gal}\times 0.10}$$

$$x=7\text{ gal}$$

299. $x\text{ gal}\times 8.34\text{ lb/gal}\times(13/100)=110\text{ gal}\times 8.34\text{ lb/gal}\times(1.2/100)$

$$x\text{ gal}=\frac{110\text{ gal}\times 8.34\text{ lb/gal}\times 0.012}{8.34\text{ lb/gal}\times 0.13}$$

$$x=10.2\text{ gal}$$

300. $6\text{ gal}\times 8.34\text{ lb/gal}\times(12/100)=x\text{ gal}\times 8.34\text{ lb/gal}\times(2/100)$

$$\frac{6\text{ gal}\times 8.34\text{ lb/gal}\times 0.12}{8.34\text{ lb/gal}\times 0.02}=x\text{ gal}$$

$$36\text{ gal}=x$$

Because 6 gal are liquid hypochlorite, a total of 36 gal – 6 gal = 30 gal water must be added.

301.

$$\frac{(50\text{ lb}\times 0.11)+(220\text{ lb}\times 0.01)}{50\text{ lb}+220\text{ lb}}\times 100$$

$$=\frac{5.5\text{ lb}+2.2\text{ lb}}{270\text{ lb}}\times 100=\frac{7.7\text{ lb}}{270\text{ lb}}\times 100=2.85\%$$

302.

$$\frac{(12\text{ gal}\times 8.34\text{ lb/gal}\times 12/100)+(60\text{ gal}\times 8.34\text{ lb/gal}\times 1.5/100)}{(12\text{ gal}\times 8.34\text{ lb/gal})+(60\text{ gal}\times 8.34\text{ lb/gal})}\times 100$$

$$=\frac{12\text{ lb}+7.5\text{ lb}}{100\text{ lb}+500\text{ lb}}\times 100=\frac{19.5\text{ lb}}{600\text{ lb}}\times 100=3.3\%$$

303. $$\frac{(16\text{ gal}\times 8.34\text{ lb/gal}\times 12/100)+(70\text{ gal}\times 8.34\text{ lb/gal}\times 1/100)}{(16\text{ gal}\times 8.34\text{ lb/gal})+(70\text{ gal}\times 8.34\text{ lb/gal})}\times 100$$

$$=\frac{16\text{ lb}+5.8\text{ lb}}{133.4\text{ lb}+583.8\text{ lb}}\times 100=\frac{21.8\text{ lb}}{717.2\text{ lb}}\times 100=3.0\%$$

304. 1,000 lb ÷ 44 lb/day = 22.7 days

305. 8 in. ÷ 12 in./ft = 0.67 ft

306. $$x=\frac{0.785\times(4\text{ ft})^2\times 3.67\text{ ft}\times 7.48\text{ gal/ft}^3}{80\text{ gpd}}=\frac{345\text{ gal}}{80\text{ gpd}}=4.3\text{ days}$$

$$\frac{24\text{ lb}}{1\text{ day}}\times\frac{1\text{ day}}{24\text{ hr}}=1\text{ lb/hr};\quad \frac{1\text{ lb}}{1\text{ hr}}\times 150\text{ hr}=150\text{ lb chlorine}$$

307. $$\frac{12\text{ lb}}{1\text{ day}}\times\frac{1\text{ day}}{24\text{ hr}}\times 111\text{ hr}=55.5\text{ lb}$$

91 lb – 55.5 lb = 35.5 lb remaining

308. 55 lb/day × 30 days = 1650 lb chlorine per month
1650 lb chlorine ÷ 150 lb/cylinder = 11 chlorine cylinders

309. $$2\text{ days}=\frac{0.785\times(3\text{ ft})^2\times x\text{ ft}\times 7.48\text{ gal/ft}^3}{52\text{ gpd}}$$

$$x=\frac{2\text{ days}\times 52\text{ gpd}}{0.785\times(3\text{ ft})^2\times 7.48\text{ gal/ft}^3}=\frac{104}{52.8}=2\text{ ft}$$

310. 1.8 mg/L + 0.9 mg/L = 2.7 mg/L

311. 23 mg/L × 0.98 MGD × 8.34 lb/gal = 18.8 lb/day

312. 60 lb/day ÷ 0.65 = 92.3 lb/day hypochlorite

313. 51 lb/day ÷ 8.34 lb/gal = 6.1 gpd

314. 3.1 mg/L = $x$ mg/L + 0.6 mg/L
3.1 mg/L – 0.6 mg/L = $x$
2.5 mg/L = $x$

315. $$\frac{30\text{ lb}\times 0.65}{(66\text{ gal}\times 8.34\text{ lb/gal})+(30\text{ lb}\times 0.65)}\times 100$$

$$=\frac{19.5\text{ lb}}{550.4\text{ lb}+19.5\text{ lb}}\times 100=\frac{19.5\text{ lb}}{569.9\text{ lb}}\times 100=3.4\%$$

316. 1620 gpm × 1440 min/day = 2,332,800 gpd
2.8 mg/L × 2.332 MGD × 8.34 lb/gal = 54.5 lb/day

317. $$2.8\text{ mg/L}\times 1.33\text{ MGD}\times 8.34\text{ lb/gal}=12{,}500\text{ mg/L}\times x\text{ MGD}\times 8.34\text{ lb/gal}$$

$$\frac{2.8\text{ mg/L}\times 1.33\text{ MGD}\times 8.34\text{ lb/gal}}{12{,}500\text{ mg/L}\times 8.34\text{ lb/gal}}=x$$

$$0.0002979\text{ MGD, or }297.9\text{ gpd}=x$$

318. $0.785 \times (0.67)^2 \times 1600 \text{ ft} \times 7.48 \text{ gal/ft}^3 = 4217 \text{ gal}$
$60 \text{ mg/L} \times 0.004217 \text{ MG} \times 8.34 \text{ lb/gal} = 2.1 \text{ lb}$

319. $x \text{ mg/L} \times 2.11 \text{ MGD} \times 8.34 \text{ lb/gal} = 3 \text{ lb/day}$

$$x = \frac{3 \text{ lb/day}}{2.11 \text{ MGD} \times 8.34 \text{ lb/gal}} = 0.17 \text{ mg/L}$$

$0.6 \text{ mg/L} - 0.5 \text{ mg/L} = 0.1$
Chlorination is assumed to be at the breakpoint.

320. $$\frac{(70 \text{ gal} \times 8.34 \text{ lb/gal} \times 12/100) + (250 \text{ gal} \times 8.34 \text{ lb/gal} \times 2/100)}{(70 \text{ gal} \times 8.34 \text{ lb/gal}) + (250 \text{ gal} \times 8.34 \text{ lb/gal})} \times 100$$

$$= \frac{70 \text{ lb} + 41.7 \text{ lb}}{584 \text{ lb} + 2085 \text{ lb}} \times 100 = \frac{117.7 \text{ lb}}{2669 \text{ lb}} \times 100 = 4.2\%$$

321. $310 \text{ lb} \div 34 \text{ lb/day} = 9.1 \text{ days}$

322. $44{,}115{,}670 \text{ gal} - 43{,}200{,}000 \text{ gal} = 915{,}670 \text{ gal}$

$x \text{ mg/L} \times 0.915 \text{ MGD} \times 8.34 \text{ lb/gal} = 18 \text{ lb/day}$

$$x = \frac{18 \text{ lb/day}}{0.915 \text{ MGD} \times 8.34 \text{ lb/gal}} = 2.4 \text{ mg/L}$$

323. $32 \text{ lb/day} \div 0.60 = 53.3 \text{ lb/day}$ hypochlorite

324. $2{,}666{,}000 \text{ gal} \div 7 \text{ days} = 380{,}857 \text{ gpd}$
$4 \text{ in.} \div 12 \text{ in./ft} = 0.33 \text{ ft}$

$$\frac{0.785 \times (4 \text{ ft})^2 \times 3.33 \text{ ft} \times 7.48 \text{ gal/ft}^3}{7 \text{ days}} = 45 \text{ gpd}$$

$x \text{ mg/L} \times 0.380 \text{ MGD} \times 8.34 \text{ lb/gal} = 20{,}000 \text{ mg/L} \times 0.000026 \times 8.34 \text{ lb/gal}$

$$x = \frac{20{,}000 \text{ mg/L} \times 0.000026 \times 8.34 \text{ lb/gal}}{0.380 \text{ MGD} \times 8.34 \text{ lb/gal}} = 1.37 \text{ mg/L}$$

325. Chlorine dose = 3.0 mg/L
$3.0 \text{ mg/L} \times 3.35 \text{ MGD} \times 8.34 \text{ lb/gal} = 83.8 \text{ lb/day}$

326. $$\frac{(12 \text{ gal} \times 12/100) + (50 \text{ gal} \times 1/100)}{12 \text{ gal} + 50 \text{ gal}} \times 100$$

$$= \frac{1.44 \text{ gal} + 0.5 \text{ gal}}{62 \text{ gal}} \times 100 = \frac{1.94 \text{ gal}}{62 \text{ gal}} = 3.1\%$$

327. $72 \text{ lb/day} = x \div 0.65$
$72 \times 0.65 = x$
$46.8 \text{ lb/day} = x$

$x \text{ mg/L} \times 1.88 \text{ MGD} \times 8.34 \text{ lb/gal} = 46.8 \text{ lb/day}$

$$x = \frac{46.8 \text{ lb/day}}{1.88 \text{ MGD} \times 8.34 \text{ lb/gal}} = 2.98 \text{ mg/L}$$

328. 400 gpm × 1440 min/day = 576,000 gpd, or 0.576 MGD

$$2.6 \text{ mg/L} \times 0.576 \text{ MGD} \times 8.34 \text{ lb/gal} = 30,000 \text{ mg/L} \times x \text{ MGD} \times 8.34 \text{ lb/gal}$$

$$\frac{2.6 \text{ mg/L} \times 0.576 \text{ MGD} \times 8.34 \text{ lb/gal}}{30,000 \text{ mg/L} \times 8.34 \text{ lb/gal}} = x \text{ MGD}$$

$$0.0000499 \text{ MGD, or } 49.9 \text{ gpd} = x$$

329. $$\frac{0.785 \times (3 \text{ ft})^2 \times 4.08 \text{ ft} \times 7.48 \text{ gal/ft}^3}{92 \text{ gpd}} = 2.3 \text{ days}$$

330. $$2 = \frac{x \text{ lb}}{(80 \text{ gal} \times 8.34 \text{ lb/gal}) + x \text{ lb}} \times 100$$

$$2 = \frac{100x}{667.2 \text{ lb} + x}$$

$$667.2 \text{ lb} + x = \frac{100x}{2}$$

$$667.2 \text{ lb} + x = 50x$$

$$667.2 \text{ lb} = 49x$$

$$\frac{667.2 \text{ lb}}{49} = x$$

$$13.7 \text{ lb} = x$$

$$\frac{13.7 \text{ lb}}{0.65} = 21.1 \text{ lb hypochlorite}$$

331. 32 lb/day ÷ 24 hr/day = 1.33 lb/hr chlorine
1.33 lb/hr × 140 hr = 186.2 lb

332. 50 lb/day × 30 days = 1500 lb
1500 lb ÷ 150 lb/cylinder = 10 cylinders required

333. 2.6% = 26,000 mg/L

334. 6600 mg/L = 0.67%

335. 29% = 290,000 mg/L

336. $$\frac{22 \text{ lb}}{1 \text{ MG} \times 8.34 \text{ lb/gal}} = \frac{22 \text{ lb}}{8.34 \text{ mil lb}} = \frac{2.64 \text{ lb}}{1 \text{ mil lb}} = 2.64 \text{ mg/L}$$

337. $$1.6 \text{ mg/L} = \frac{1.6 \text{ lb}}{1 \text{ mil lb}} = \frac{1.6 \text{ lb}}{1 \text{ mil lb}/8.34 \text{ lb/gal}} = \frac{1.6 \text{ lb}}{0.12 \text{ MG}} = 13.3 \text{ lb/MG}$$

338. $$\frac{25 \text{ lb}}{1 \text{ MG} \times 8.34 \text{ mL/lb}} = \frac{25 \text{ lb}}{8.34 \text{ mil lb}} = \frac{2.99 \text{ lb}}{1 \text{ mil lb}} = 2.99 \text{ mg/L}$$

339. H molecular weight = 2 atoms × 1.008 atomic weight = 2.016
Si molecular weight = 1 atom × 28.06 atomic weight = 28.06
F molecular weight = 6 atoms × 19.00 atomic weight = 114.00
Molecular weight of $H_2SiF_6$ = 144.076
(114.00 ÷ 144.076) × 100 = 79.1%

340. Na molecular weight = 1 atom × 22.997 atomic weight = 22.997
F molecular weight = 1 atom × 19.00 atomic weight = 19.00
Molecular weight of NaF = 41.997
(19.00 ÷ 41.997) × 100 = 45.2%

341. $$\frac{1.6\text{ mg/L} \times 0.98\text{ MG} \times 8.34\text{ lb/gal}}{(98/100) \times (60.6/100)} = \frac{13.1}{0.98 \times 0.606} = \frac{13.1}{0.59} = 22\text{ lb/day}$$

342. $$\frac{1.4\text{ mg/L} \times 1.78\text{ MGD} \times 8.34\text{ lb/gal}}{\frac{98}{100} \times \frac{60.6}{100}} = \frac{20.8}{0.98 \times 0.606} = \frac{20.8}{0.59} = 35.3\text{ lb/day}$$

343. $$\frac{1.4\text{ mg/L} \times 2.880\text{ MGD} \times 8.34\text{ lb/gal}}{0.8\text{ lb}} = 42.1\text{ lb/day}$$

344. $$\frac{1.1\text{ mg/L} \times 3.08\text{ MGD} \times 8.34\text{ lb/gal}}{0.45\text{ lb}} = 62.8\text{ lb/day}$$

345. 1.2 mg/L – 0.08 mg/L = 1.13

$$\frac{1.13\text{ mg/L} \times 0.810\text{ MGD} \times 8.34\text{ lb/gal}}{0.45\text{ lb}} = 169\text{ lb/day}$$

346. $$\frac{91\text{ lb} \times (98/100)}{(55\text{ gal} \times 8.34\text{ lb/gal}) + (9\text{ lb} \times 98/100)} \times 100 = \frac{8.82\text{ lb}}{459\text{ lb} + 8.82\text{ lb}} \times 100 = 1.9\%$$

347. $$\frac{20\text{ lb}}{(80\text{ gal} \times 8.34\text{ lb/gal}) + 20\text{ lb}} \times 100$$

$$= \frac{20\text{ lb}}{667.2\text{ lb} + 20\text{ lb}} \times 100 = \frac{20}{687.2} \times 100 = 2.9\%$$

348. $$1.4 = \frac{x\text{ lb} \times (98/100)}{(220\text{ gal} \times 8.34\text{ lb/gal}) + (x\text{ lb} \times 98/100)} \times 100$$

$$1.4 = \frac{0.98x \times 100}{1835 + 0.98x}$$

$$1.4 = \frac{98x}{1835 + 0.98x}$$

$$1.4 \times (1835 + 0.98x) = 98x$$

$$2569 + 1.37x = 98x$$

$$2569 = 96.63x$$

$$\frac{2569}{96.63} = x$$

$$26.6\text{ lb} = x$$

349. $$\frac{11\text{ lb} \times (98/100)}{(60\text{ gal} \times 8.34\text{ lb/gal}) + (11\text{ lb} \times 98/100)} \times 100 = \frac{10.78\text{ lb}}{500\ \text{ lb} + 10.8\text{ lb}} \times 100 = 2.1\%$$

350. $$3 = \frac{x \text{ lb} \times (98/100)}{(160 \text{ gal} \times 8.34 \text{ lb/gal}) + (x \text{ lb} \times 98/100)} \times 100$$

$$3 = \frac{0.98x \times 100}{1334 + 0.98x}$$

$$3 = \frac{98x}{1334 + 0.98x}$$

$$3 \times (1334 + 0.98x) = 98x$$

$$4002 + 2.94x = 98x$$

$$4002 = 98x - 2.94x$$

$$4002 = 95.09x$$

$$\frac{4002}{95.09} = x$$

$$42 \text{ lb} = x$$

351. $$1.2 \text{ mg/L} \times 4.23 \text{ MGD} \times 8.34 \text{ lb/gal} = 240{,}000 \text{ mg/L} \times x \text{ MGD} \times 8.34 \text{ lb/gal} \times 1.2 \times (80/100)$$

$$\frac{1.2 \text{ mg/L} \times 4.23 \text{ MGD} \times 8.34 \text{ lb/gal}}{240{,}000 \text{ mg/L} \times 8.34 \text{ lb/gal} \times 1.2 \times 0.80} = x \text{ MGD}$$

$$0.0000217 \text{ MGD, or } 21.7 \text{ gpd} = x$$

352. $$1.2 \text{ mg/L} \times 3.1 \text{ MGD} \times 8.34 \text{ lb/gal} = 220{,}000 \text{ mg/L} \times x \text{ MGD} \times 9.7 \text{ lb/gal} \times (80/100)$$

$$\frac{1.2 \text{ mg/L} \times 3.1 \text{ MGD} \times 8.34 \text{ lb/gal}}{220{,}000 \text{ mg/L} \times 9.7 \text{ lb/gal} \times 1.2 \times 0.80} = x \text{ MGD}$$

$$0.0000199 \text{ MGD, or } 19.9 \text{ gpd} = x$$

353. 1.8 mg/L – 0.09 mg/L = 1.71 mg/L

$$1.71 \text{ mg/L} \times 0.91 \text{ MGD} \times 8.34 \text{ lb/gal} = 22{,}000 \text{ mg/L} \times x \text{ MGD} \times 8.34 \text{ lb/gal} \times (46.10/100)$$

$$\frac{1.71 \text{ mg/L} \times 0.91 \text{ MGD} \times 8.34 \text{ lb/gal}}{22{,}000 \text{ mg/L} \times 8.34 \text{ lb/gal} \times 0.4610} = x \text{ MGD}$$

$$0.0001543 \text{ MGD, or } 154.3 \text{ gpd} = x$$

354. $$1.6 \text{ mg/L} \times 1.52 \text{ MGD} \times 8.34 \text{ lb/gal} = 24{,}000 \text{ mg/L} \times x \text{ MGD} \times 8.34 \text{ lb/gal} \times (45.25/100)$$

$$\frac{1.6 \text{ mg/L} \times 1.52 \text{ MGD} \times 8.34 \text{ lb/gal}}{24{,}000 \text{ mg/L} \times 8.34 \text{ lb/gal} \times 0.4575} = x \text{ MGD}$$

$$0.0002239 \text{ MGD, or } 223.9 \text{ gpd} = x$$

355. $$\frac{80 \text{ gpd} \times 3785 \text{ mL/gal}}{1440 \text{ min/day}} = 210.3 \text{ mL/min}$$

356. $$1.0 \text{ mg/L} \times 2.78 \text{ MGD} \times 8.34 \text{ lb/gal} = 250{,}000 \text{ mg/L} \times x \text{ MGD} \times 9.8 \text{ lb/gal} \times (80/100)$$

$$\frac{1.0 \text{ mg/L} \times 2.78 \text{ MGD} \times 8.34 \text{ lb/gal}}{250{,}000 \text{ mg/L} \times 9.8 \text{ lb/gal} \times 0.80} = x \text{ MGD}$$

$$0.0000118 \text{ MGD, or } 11.8 \text{ gpd} = x$$

$$\frac{11.8 \text{ gpd} \times 3785 \text{ mL/gal}}{1440 \text{ min/day}} = 31 \text{ mL/min}$$

357. $$\frac{x \text{ mg/L} \times 1.52 \text{ MGD} \times 8.34 \text{ lb/gal}}{(98/100) \times (61/100)} = 40 \text{ lb/day}$$

$$x \text{ mg/L} \times 1.52 \text{ MGD} \times 8.34 \text{ lb/gal} = 40 \text{ lb/day} \times 0.98 \times 0.61$$

$$x \text{ mg/L} = \frac{40 \text{ lb/day} \times 0.98 \times 0.61}{1.52 \text{ MGD} \times 8.34 \text{ lb/gal}}$$

$$x \text{ mg/L} = \frac{23.9}{12.7}$$

$$x = 1.89 \text{ mg/L}$$

358. $$\frac{x \text{ mg/L} \times 0.33 \text{ MGD} \times 8.34 \text{ lb/gal}}{(98/100) \times (45.25/100)} = 6 \text{ lb/day}$$

$$x \text{ mg/L} \times 0.33 \text{ MGD} \times 8.34 \text{ lb/gal} = 6 \text{ lb/day} \times 0.98 \times 0.4525$$

$$x \text{ mg/L} = \frac{6 \text{ lb/day} \times 0.98 \times 0.4525}{0.33 \text{ MGD} \times 8.34 \text{ lb/gal}}$$

$$x \text{ mg/L} = \frac{2.7}{2.8}$$

$$x = 0.97 \text{ mg/L}$$

359. $$x \text{ mg/L} \times 3.85 \text{ MGD} \times 8.34 \text{ lb/gal} = 200{,}000 \text{ mg/L} \times 0.000032 \text{ MGD} \times 9.8 \text{ lb/gal} \times (80/100)$$

$$x \text{ mg/L} = \frac{200{,}000 \text{ mg/L} \times 0.000032 \text{ MGD} \times 9.8 \text{ lb/gal} \times 0.80}{3.85 \text{ MGD} \times 8.34 \text{ lb/gal}}$$

$$x \text{ mg/L} = \frac{50.18}{32.11}$$

$$x = 1.6 \text{ mg/L}$$

360. $x \text{ mg/L} \times 1.92 \text{ MGD} \times 8.34 \text{ lb/gal} = 110{,}000 \text{ mg/L} \times 0.000028 \text{ MGD} \times 9.1 \text{ lb/gal} \times (80/100)$

$$x \text{ mg/L} = \frac{110{,}000 \text{ mg/L} \times 0.000028 \text{ MGD} \times 9.10 \text{ lb/gal} \times 0.80}{1.92 \text{ MGD} \times 8.34 \text{ lb/gal}}$$

$$x = 1.4 \text{ mg/L}$$

361. $x \text{ mg/L} \times 2.73 \text{ MGD} \times 8.34 \text{ lb/gal} = 30{,}000 \text{ mg/L} \times 0.000110 \text{ MGD} \times 8.34 \text{ lb/gal} \times (45.25/100)$

$$x \text{ mg/L} = \frac{30{,}000 \text{ mg/L} \times 0.000110 \text{ MGD} \times 8.34 \text{ lb/gal} \times 0.4525}{2.73 \text{ MGD} \times 8.34 \text{ lb/gal}}$$

$$x = 0.55 \text{ mg/L}$$

362. $\dfrac{(600 \text{ lb} \times 15/100) + (2600 \text{ lb} \times 25/100)}{600 \text{ lb} + 2600 \text{ lb}} \times 100 = \dfrac{90 \text{ lb} + 650 \text{ lb}}{3200 \text{ lb}} \times 100 = 23\%$

363. $(900 \text{ lb} \times 25/100) + (300 \text{ lb} \times 15/100) = 1200 \text{ lb} \times (x/100)$

$$225 \text{ lb} + 45 \text{ lb} = 1200 \text{ lb} \times (x/100)$$

$$270 = 12x$$

$$\frac{270}{12} = x$$

$$22.5 = x$$

364. $\dfrac{(400 \text{ gal} \times 9.4 \text{ lb/gal} \times 16/100) + (2200 \text{ lb} \times 9.10 \text{ lb/gal} \times 26/100)}{(400 \text{ gal} \times 19.4 \text{ lb/gal}) + (2200 \text{ lb} \times 9.10 \text{ lb/gal})} \times 100$

$$= \frac{601.6 \text{ lb} + 4404.4 \text{ lb}}{3760 \text{ lb} + 20{,}020 \text{ lb}} \times 100 = \frac{5006 \text{ lb}}{23{,}780 \text{ lb}} \times 100 = 21.1\%$$

365. $\dfrac{(325 \text{ gal} \times 9.06 \text{ lb/gal} \times 11/100) + (1100 \text{ lb} \times 9.8 \text{ lb/gal} \times 20/100)}{(325 \text{ gal} \times 9.06 \text{ lb/gal}) + (1100 \text{ lb} \times 9.8 \text{ lb/gal})} \times 100$

$$= \frac{324 \text{ lb} + 2156 \text{ lb}}{2944.5 \text{ lb} + 10{,}780 \text{ lb}} \times 100 = \frac{2480 \text{ lb}}{13{,}724.5 \text{ lb}} \times 100 = 18.1\%$$

366. Density = 8.34 lb/gal × 1.075 = 8.97

$$\frac{(220 \text{ gal} \times 8.97 \text{ lb/gal} \times 10/100) + (1600 \text{ lb} \times 9.5 \text{ lb/gal} \times 15/100)}{(220 \text{ gal} \times 8.97 \text{ lb/gal}) + (1600 \text{ lb} \times 9.5 \text{ lb/gal})} \times 100$$

$$= \frac{197.3 \text{ lb} + 2280 \text{ lb}}{1973 \text{ lb} + 15{,}200 \text{ lb}} \times 100 = \frac{2477.3 \text{ lb}}{17{,}173 \text{ lb}} \times 100 = 14.4\%$$

367. 2.9000 = 29,000 mg/L

368. H molecular weight = 2 atoms × 1.008 atomic weight = 2.016
Si molecular weight = 1 atom × 28.06 atomic weight = 28.06
F molecular weight = 6 atoms × 19.00 atomic weight = 114.00
Molecular weight of $H_2SiF_6$ = 144.076
(114.00 ÷ 144.076) × 100 = 79.1%

369. $$\frac{27\text{ lb}}{1\text{ MG}\times 8.34\text{ lb/gal}}=\frac{27\text{ lb}}{8.34\text{ mil lb}}\times\frac{3.24}{1\text{mil lb}}=3.24\text{ mg/L}$$

370. $$\frac{1.6\text{ mg/L}\times 2.111\text{ MGD}\times 8.34\text{ lb/gal}}{(98/100)\times(61.2/100)}=47\text{ lb/day}$$

371. Molecular weight of NaF is 41.997.
(19.00 ÷ 41.997) × 100 = 45.2%

372. $$\frac{80\text{ lb}\times 98/100}{(600\text{ gal}\times 8.34\text{ lb/gal})+(80\text{ lb}\times 98/100)}\times 100$$

$$=\frac{78.4\text{ lb}}{5004\text{ lb}+78.4\text{ lb}}\times 100=\frac{78.4\text{ lb}}{5082.4\text{ lb}}\times 100=1.5\%$$

373. 28,000 mg/L = 2.8%

374. $$\frac{80\text{ gpd}\times 3785\text{ mL/gal}}{1440\text{ min/day}}=210\text{ mL/min}$$

375. $$\frac{1.5\text{ mg/L}\times 2.45\text{ MGD}\times 8.34\text{ lb/gal}}{(98/100)\times(45.25/100)}=156\text{ lb/day}$$

376. $$3=\frac{x\text{ lb}\times(98/100)}{(600\text{ gal}\times 8.34\text{ lb/gal})+(x\text{ lb}\times 98/100)}\times 100$$

$$3=\frac{0.98x\times 100}{5004\text{ lb}+0.98x}$$

$$3=\frac{98x}{5004\text{ lb}+0.98x}$$

$$3\times(5004\text{ lb}+0.98x)=98x$$

$$15{,}012\text{ lb}+2.94x=98x$$

$$15{,}012\text{ lb}=95.06x$$

$$158\text{ lb}=x$$

377. $$1.4\text{ mg/L}\times 4.11\text{ MGD}\times 8.34\text{ lb/gal}=210{,}000\text{ mg/L}\times x\text{ MGD}\times 8.34\text{ lb/gal}\times 1.3\times(80/100)$$

$$\frac{1.4\text{ mg/L}\times 4.11\text{ MGD}\times 8.34\text{ lb/gal}}{210{,}000\text{ mg/L}\times 8.34\text{ lb/gal}\times 1.3\times 0.80}=x\text{ MGD}$$

$$0.0000262\text{ MGD, or }26.2\text{ gpd}=x$$

378. $$\frac{30\text{ lb}\times 98/100}{(140\text{ gal}\times 8.34\text{ lb/gal})+(30\text{ lb}\times 98/100)}\times 100$$

$$=\frac{29.4\text{ lb}}{1167.8\text{ lb}+29.4\text{ lb}}\times 100=\frac{29.4\text{ lb}}{1197.2\text{ lb}}\times 100=2.45\%$$

379. 1.4 mg/L – 0.09 mg/L = 1.31 mg/L

$$\frac{1.31 \text{ mg/L} \times 1.88 \text{ MGD} \times 8.34 \text{ lb/gal}}{0.44} = 46.7 \text{ lb/day}$$

380. $$1.3 \text{ mg/L} \times 2.8 \text{ MGD} \times 8.34 \text{ lb/gal} = 200{,}000 \text{ mg/L} \times x \text{ MGD} \times 9.8 \text{ lb/gal} \times (80/100)$$

$$\frac{1.3 \text{ mg/L} \times 2.8 \text{ MGD} \times 8.34 \text{ lb/gal}}{200{,}000 \text{ mg/L} \times 9.8 \text{ lb/gal} \times 0.80} = x \text{ MGD}$$

$$0.0000193 \text{ MGD, or } 19.3 \text{ gpd} = x$$

381. $$\frac{(500 \text{ lb} \times 15/100) + (1600 \text{ lb} \times 20/100)}{500 \text{ lb} + 1600 \text{ lb}} \times 100$$

$$= \frac{75 \text{ lb} + 320 \text{ lb}}{2100 \text{ lb}} \times 100 = \frac{395 \text{ lb}}{2100 \text{ lb}} \times 100 = 18.8\%$$

382. $$\frac{x \text{ mg/L} \times 1.10 \text{ MGD} \times 8.34 \text{ lb/gal}}{(98/100) \times (61.1/100)} = 41 \text{ lb/day}$$

$$x \text{ mg/L} \times 1.10 \text{ MGD} \times 8.34 \text{ lb/gal} = 41 \text{ lb/day} \times 0.98 \times 0.611$$

$$x = \frac{41 \text{ lb/day} \times 0.98 \times 0.611}{1.10 \text{ MGD} \times 8.34 \text{ lb/gal}} = 2.7 \text{ mg/L}$$

383. 1400 gpm × 1440 min/day = 2,016,000 gpd = 2.016 MGD

$$x \text{ mg/L} \times 2.016 \text{ MGD} \times 8.34 \text{ lb/gal} = 110{,}000 \text{ mg/L} \times 0.000040 \text{ MGD} \times 9.14 \text{ lb/gal} \times (80/100)$$

$$x = \frac{110{,}000 \text{ mg/L} \times 0.000040 \text{ MGD} \times 9.14 \text{ lb/gal} \times 0.8}{2.016 \text{ MGD} \times 8.34 \text{ lb/gal}} = 1.92 \text{ mg/L}$$

384. $$\frac{(235 \text{ gal} \times 9.14 \text{ lb/gal} \times 10/100) + (600 \text{ gal} \times 9.8 \text{ lb/gal} \times 20/100)}{(235 \text{ gal} \times 9.14 \text{ lb/gal}) + (600 \text{ gal} \times 9.8 \text{ lb/gal})} \times 100$$

$$= \frac{215 \text{ lb} + 1176 \text{ lb}}{2147.9 \text{ lb} + 5880 \text{ lb}} \times 100 = \frac{1391 \text{ lb}}{8027.9 \text{ lb}} \times 100 = 17.3\%$$

385. $$1.1 \text{ mg/L} \times 2.88 \text{ MGD} \times 8.34 \text{ lb/gal} = 200{,}000 \text{ mg/L} \times x \text{ MGD} \times 9.8 \text{ lb/gal} \times (80/100)$$

$$\frac{1.1 \text{ mg/L} \times 2.88 \text{ MGD} \times 8.34 \text{ lb/gal}}{200{,}000 \text{ mg/L} \times 9.8 \text{ lb/gal} \times 0.80} = x \text{ MGD}$$

$$0.0000168 \text{ MGD, or } 16.8 \text{ gpd} = x$$

$$\frac{16.8 \text{ gpd} \times 3785 \text{ mL/gal}}{1440 \text{ min/day}} = 44.2 \text{ mL/min}$$

386. Density = 8.34 lb/gal × 1.115 = 9.3 lb/gal

$$\frac{(131\ \text{gal} \times 9.3\ \text{lb/gal} \times 9/100) + (900\ \text{gal} \times 9.4\ \text{lb/gal} \times 15/100)}{(131\ \text{gal} \times 9.3\ \text{lb/gal}) + (900\ \text{gal} \times 9.4\ \text{lb/gal})} \times 100$$

$$= \frac{109.6\ \text{lb} + 1269\ \text{lb}}{1218\ \text{lb} + 8460\ \text{lb}} \times 100 = \frac{1378.6\ \text{lb}}{9678\ \text{lb}} \times 100 = 14.2\%$$

387. $$x\ \text{mg/L} \times 2.90\ \text{MGD} \times 8.34\ \text{lb/gal} = 50{,}000\ \text{mg/L} \times 0.000120\ \text{MGD} \times 8.34\ \text{lb/gal} \times (45.25/100)$$

$$x = \frac{50{,}000\ \text{mg/L} \times 0.000120\ \text{MGD} \times 8.34\ \text{lb/gal} \times 0.4525}{2.90\ \text{MGD} \times 8.34\ \text{lb/gal}} = 0.94\ \text{mg/L}$$

0.94 mg/L added + 0.2 mg/L in raw water = 0.96 mg/L in finished water

388. $$\frac{x\ \text{mg/L}}{50.045} = \frac{39\ \text{mg/L}}{20.04}$$

$$x\ \text{mg/L} = \frac{50.045 \times 39\ \text{mg/L}}{20.04}$$

$$x = 98.2\ \text{mg/L}$$

389. $$\frac{x\ \text{mg/L}}{50.045} = \frac{33\ \text{mg/L}}{12.16}$$

$$x\ \text{mg/L} = \frac{50.045 \times 33\ \text{mg/L}}{12.16}$$

$$x = 136\ \text{mg/L}$$

390. $$\frac{x\ \text{mg/L}}{50.045} = \frac{18\ \text{mg/L}}{20.04}$$

$$x\ \text{mg/L} = \frac{50.045 \times 18\ \text{mg/L}}{20.04}$$

$$x = 44.9\ \text{mg/L}$$

391. 75 mg/L + 91 mg/L = 166 mg/L as $CaCO_3$

392. $$\frac{x\ \text{mg/L}}{50.045} = \frac{30\ \text{mg/L}}{20.04}$$

$$x\ \text{mg/L} = \frac{50.045 \times 30\ \text{mg/L}}{20.04}$$

$x = 74.9$ mg/L calcium as $CaCO_3$

$$\frac{x\ \text{mg/L}}{50.045} = \frac{10\ \text{mg/L}}{12.16}$$

$$x\ \text{mg/L} = \frac{50.045 \times 10\ \text{mg/L}}{12.16}$$

$x = 41$ mg/L magnesium ast $CaCO_3$

74.9 mg/L + 41 mg/L = 115.9 mg/L total hardness

393. $$\frac{x\text{ mg/L}}{50.045} = \frac{21\text{ mg/L}}{20.04}$$

$$x\text{ mg/L} = \frac{50.045 \times 21\text{ mg/L}}{20.04}$$

$x = 52.4$ mg/L calcium as $CaCO_3$

$$\frac{x\text{ mg/L}}{50.045} = \frac{15\text{ mg/L}}{12.16}$$

$$x\text{ mg/L} = \frac{50.045 \times 15\text{ mg/L}}{12.16}$$

$x = 61.7$ mg/L magnesium as $CaCO_3$

52.4 mg/L + 61.7 mg/L = 114.1 mg/L total hardness

394. Total hardness = carbonate hardness = 121 mg/L. There is no noncarbonate hardness in this water.

395. 122 mg/L = 105 mg/L + $x$ mg/L
122 mg/L – 105 mg/L = $x$ mg/L
17 mg/L noncarbonate hardness = $x$
Carbonate hardness = 105 mg/L

396. 116 mg/L = 91 mg/L + $x$ mg/L
116 mg/L – 91 mg/L = $x$ mg/L
25 mg/L noncarbonate hardness = $x$
Carbonate hardness = 91 mg/L

397. Alkalinity is greater than total hardness; therefore, all of the hardness is carbonate hardness = 99 mg/L as $CaCO_3$.

398. 121 mg/L = 103 mg/L + $x$ mg/L
121 mg/L – 103 mg/L = $x$ mg/L
18 mg/L noncarbonate hardness = $x$
Carbonate hardness = 103 mg/L

399. $$\frac{A \times N \times 50{,}000}{\text{Sample (mL)}} = \frac{2.0\text{ mL} \times 0.02\ N \times 50{,}000}{100\text{ mL}} = 20\text{ mg/L as }CaCO_3$$

400. $$\frac{A \times N \times 50{,}000}{\text{Sample (mL)}} = \frac{1.4\text{ mL} \times 0.02\ N \times 50{,}000}{100\text{ mL}} = 14\text{ mg/L as }CaCO_3$$

401. $$\frac{A \times N \times 50{,}000}{\text{Sample (mL)}} = \frac{0.3\text{ mL} \times 0.02\ N \times 50{,}000}{100\text{ mL}} = 3\text{ mg/L}$$

$$\text{Total alkalinity} = \frac{6.7\text{ mL} \times 0.02\ N \times 50{,}000}{100\text{ mL}} = 67\text{ mg/L}$$

402. Phenolphthalein alkalinity = 0 mg/L

$$\text{Total alkalinity} = \frac{6.9\text{ mL} \times 0.02\ N \times 50{,}000}{100\text{ mL}} = 69\text{ mg/L}$$

403. $\dfrac{A \times N \times 50{,}000}{\text{Sample (mL)}} = \dfrac{0.5\ \text{mL} \times 0.02\ N \times 50{,}000}{100\ \text{mL}} = 5\ \text{mg/L}$

$$\text{Total alkalinity} = \frac{5.7\ \text{mL} \times 0.02\ N \times 50{,}000}{100\ \text{mL}} = 57\ \text{mg/L}$$

404. Bicarbonate alkalinity = T – 2P = 51 mg/L – 2(8 mg/L) = 35 mg/L as $CaCO_3$
Carbonate alkalinity = 2P = 2(8 mg/L) = 16 mg/L as $CaCO_3$
Hydroxide alkalinity = 0 mg/L

405. Bicarbonate alkalinity = T = 67 mg/L as $CaCO_3$
Carbonate alkalinity = 0 mg/L
Hydroxide alkalinity = 0 mg/L

406. Bicarbonate alkalinity = 0 mg/L
Carbonate alkalinity = 2T – 2P = 2(23 mg/L) – 2(12 mg/L) = 22 mg/L as $CaCO_3$
Hydroxide alkalinity = 2P – T = 2(12 mg/L) – 23 mg/L = 1 mg/L as $CaCO_3$

407. $\text{Phenolphthalein alkalinity} = \dfrac{A \times N \times 50{,}000}{\text{Sample (mL)}} = \dfrac{1.3\ \text{mL} \times 0.02\ N \times 50{,}000}{100\ \text{mL}}$

$$= 13\ \text{mg/L as } CaCO_3$$

$$\text{Total alkalinity} = \frac{5.3\ \text{mL} \times 0.02\ N \times 50{,}000}{100\ \text{mL}} = 53\ \text{mg/L as } CaCO_3$$

Bicarbonate alkalinity = T – 2P = 51 mg/L – 2(13 mg/L) = 25 mg/L as $CaCO_3$
Carbonate alkalinity = 2P = 2(13 mg/L) = 26 mg/L as $CaCO_3$
Hydroxide alkalinity = 0 mg/L

408. $\text{Phenolphthalein alkalinity} = \dfrac{A \times N \times 50{,}000}{\text{Sample (mL)}} = \dfrac{1.5\ \text{mL} \times 0.02\ N \times 50{,}000}{100\ \text{mL}}$

$$= 15\ \text{mg/L as } CaCO_3$$

$$\text{Total alkalinity} = \frac{2.9\ \text{mL} \times 0.02\ N \times 50{,}000}{100\ \text{mL}} = 29\ \text{mg/L as } CaCO_3$$

From the alkalinity table:
Bicarbonate alkalinity = 0 mg/L
Carbonate alkalinity = 2P = 2(15 mg/L) = 30 mg/L as $CaCO_3$
Hydroxide alkalinity = 0 mg/L

409. $A = CO_2$ (mg/L) × (56/44) = 8 mg/L × (56/44) = 10 mg/L
$B$ = Alkalinity (mg/L) × (56/100) = 130 mg/L × (56/100) = 73 mg/L
$C$ = 0 mg/L
$D = Mg^{2+}$ (mg/L) × (56/24.3) = 22 mg/L × (56/24.3) = 51 mg/L

$$\text{Quick lime dosage} = \frac{(10\ \text{mg/L} + 73\ \text{mg/L} + 0 + 51\ \text{mg/L}) \times 1.15}{0.90}$$

$$= \frac{134\ \text{mg/L} \times 1.15}{0.90} = 171\ \text{mg/L as CaO}$$

410. $A = CO_2$ (mg/L) × (74/44) = 5 mg/L × (74/44) = 8 mg/L
$B$ = Alkalinity (mg/L) × (74/100) = 164 mg/L × (74/100) = 121 mg/L
$C$ = 0 mg/L
$D = Mg^{2+}$ (mg/L) × (74/24.3) = 17 mg/L × (74/24.3) = 52 mg/L

$$\text{Hydrated lime dosage} = \frac{(8\text{ mg/L} + 121\text{ mg/L} + 0 + 52\text{ mg/L}) \times 1.15}{0.90}$$

$$= \frac{181\text{ mg/L} \times 1.15}{0.90} = 231\text{ mg/L as Ca(OH)}_2$$

411. $A = CO_2$ (mg/L) × (74/44) = 6 mg/L × (74/44) = 10 mg/L
$B$ = Alkalinity (mg/L) × (24/100) = 110 mg/L × (74/100) = 81 mg/L
$C$ = 0 mg/L
$D = Mg^{2+}$ (mg/L) × (74/24.3) = 12 mg/L × (74/24.3) = 37 mg/L

$$\text{Hydrated lime dosage} = \frac{(10\text{ mg/L} + 81\text{ mg/L} + 0 + 37\text{ mg/L}) \times 1.15}{0.90}$$

$$= \frac{128\text{ mg/L} \times 1.15}{0.90} = 164\text{ mg/L as Ca(OH)}_2$$

412. $A$ = Carbon dioxide (mg/L) × (56/44) = 9 mg/L × (56/44) = 11 mg/L
$B$ = Alkalinity (mg/L) × (56/100) = 180 mg/L × (56/100) = 101 mg/L
$C$ = 0 mg/L
$D = Mg^{2+}$ (mg/L) × (56/24.3) = 18 mg/L × (56/24.3) = 41 mg/L

$$\text{Quick lime dosage} = \frac{(11\text{ mg/L} + 101\text{ mg/L} + 0 + 41\text{ mg/L}) \times 1.15}{0.90}$$

$$= \frac{153\text{ mg/L} \times 1.15}{0.90} = 196\text{ mg/L as CaO}$$

413. 260 mg/L – 169 mg/L = 91 mg/L
Soda ash = 91 mg/L × (106/100) = 96 mg/L
414. 240 mg/L – 111 mg/L = 129 mg/L
Soda ash = 129 mg/L × (106/100) = 137 mg/L
415. 264 mg/L – 170 mg/L = 94 mg/L
Soda ash = 94 mg/L × (106/100) = 100 mg/L
416. 228 mg/L – 108 mg/L = 120 mg/L
Soda ash = 120 mg/L × (106/100) = 127 mg/L
417. Excess lime = (8 mg/L + 130 mg/L + 0 + 66 mg/L) × 0.15
= 204 mg/L × 0.15 = 31 mg/L
Total carbon dioxide dosage = (31 mg/L × 44/100) + (4 mg/L × 44/24.3)
= 14 mg/L + 7 mg/L = 21 mg/L
418. Excess lime = (8 mg/L + 90 mg/L + 7 + 109 mg/L) × 0.15
= 213 mg/L × 0.15 = 32 mg/L
Total carbon dioxide = (32 mg/L × 44/74) + (3 mg/L × 44/24.3)
= 19 mg/L + 5.4 mg/L = 24.4 mg/L
419. Excess lime = (7 mg/L + 109 mg/L + 3 mg/L + 52 mg/L) × 0.15
= 171 mg/L × 0.15 = 26 mg/L
Total carbon dioxide = (26 mg/L × 44/74) + (5 mg/L × 44/24.3)
= 15.5 mg/L + 9 mg/L = 24.5 mg/L

420. Excess lime = (6 mg/L + 112 mg/L + 6 mg/L + 45 mg/L) × 0.15
= 169 mg/L × 0.15 = 26 mg/L
Total carbon dioxide = (26 mg/L × 44/74) + (4 mg/L × 44/24.3)
= 15 mg/L + 7 mg/L = 22 mg/L
421. 200 mg/L × 2.47 MGD × 8.34 lb/gal = 4120 lb/day
422. 180 mg/L × 3.12 MGD × 8.34 lb/gal = 4684 lb/day
4684 lb/day ÷ 1440 min/day = 3.3 lb/min
423. 60 mg/L × 4.20 MGD × 8.34 lb/gal = 2102 lb/day
2102 lb/day ÷ 24 hr/day = 87.5 lb/hr
424. 130 mg/L × 1.85 MGD × 8.34 lb/gal = 2006 lb/day
2006 lb/day ÷ 1440 min/day = 1.4 lb/min
425. 40 mg/L × 3.11 MGD × 8.34 lb/gal = 1038 lb/day
1038 lb/day ÷ 24 hr/day = 43 lb/hr
43 lb/hr ÷ 60 min/hr = 0.7 lb/min
426. 211 mg/L ÷ 17.12 mg/L/gr/gal = 12.3 gr/gal
427. 12.3 mg/L × 17.12 mg/L/gr/gal) = 211 mg/L
428. 240 mg/L ÷ 17.12 mg/L/gpd = 14 gr/gal
429. 14 gpg × 17.12 mg/L/gr/gal) = 240 mg/L
430. 25,000 gr/ft$^3$ × 105 ft$^3$ = 2,625,000 gr
431. 0.785 × (6 ft)$^2$ × 4.17 ft = 119 ft$^3$
25,000 gr/ft$^3$ × 119 ft$^3$ = 2,975,000 gr
432. 20,000 gr/ft$^3$ × 260 ft$^3$ = 5,200,000 gr
433. 0.785 × (8ft)$^2$ × 5 ft = 251 ft$^3$
22,000 gr/ft$^3$ × 251 ft$^3$ = 5,522,000 gr
434. 2,210,000 gr ÷ 18.1 gr/gal = 122,099 gal
435. 4,200,000 gr ÷ 16.0 gr/gal = 262,500 gal
436. 270 mg/L ÷ 17.12 mg/L/gr/gal = 15.8 gpg
3,650,000 gr ÷ 15.8 gr/gal = 231,013 gal
437. 21,000 gr/ft$^3$ × 165 ft$^3$ = 3,465,000 gr
3,465,000 gr ÷ 14.6 gr/gal = 237,329 gal
438. 0.785 × (3 ft)$^2$ × 2.6 ft = 18.4 ft$^3$
22,000 gr/ft$^3$ × 18.4 ft$^3$ = 404,800 gr
404,800 gr ÷ 18.4 gr/gal = 22,000 gal
439. 575,000 gal ÷ 25,200 gph = 22.8 hr
440. 766,000 gal ÷ 26,000 gph = 29.5 hr
441. 230 gpm × 60 min/hr = 13,800 gph
348,000 gal ÷ 13,800 gph = 25.2 hr
442. 3,120,000 gr ÷ 14 gpg = 222,857 gal
200 gpm × 60 min/hr = 12,000 gph
222,857 gal ÷ 12,000 gph = 18.6 hr
443. 3,820,000 gr ÷ 11.6 gpg = 329,310 gal
290,000 gpd ÷ 24 hr/day = 12,083 gph
329,310 gal ÷ 12,083 gph = 27.3 hr

444. $\frac{\text{0.5 lb salt} \times \text{2300 Kgr}}{\text{Kilograins remaining}} = \text{1150 lb salt required}$

445. $\frac{0.4 \text{ lb salt} \times 1330 \text{ Kgr}}{\text{Kilograins remaining}} = 532 \text{ lb salt required}$

446. $\frac{410 \text{ lb salt}}{1.19 \text{ lb salt per gal brine}} = 345 \text{ gal of 13\% brine}$

447. $\frac{420 \text{ lb salt}}{1.29 \text{ lb salt per gal brine}} = 326 \text{ gal of 14\% brine}$

448. $\frac{0.5 \text{ lb salt} \times 1310 \text{ Kgr}}{\text{Kilograins remaining}} = 655 \text{ lb salt required}$

$$\frac{655 \text{ lb salt}}{1.09 \text{ lb salt per gal brine}} = 601 \text{ gal of 12\% brine}$$

449. $\frac{x \text{ mg/L}}{50.045} = \frac{44 \text{ mg/L}}{20.04}$

$$x \text{ mg/L} = \frac{50.045 \times 44 \text{ mg/L}}{20.04}$$

$$x = 110 \text{ mg/L as } CaCO_3$$

450. $\frac{1.8 \text{ mL} \times 0.02\ N \times 50{,}000}{100 \text{ mL}} = 18 \text{ mg/L as } CaCO_3$

451. $\frac{x \text{ mg/L}}{50.045} = \frac{31 \text{ mg/L}}{12.16}$

$$x \text{ mg/L} = \frac{50.045 \times 31 \text{ mg/L}}{12.16}$$

$$x = 128 \text{ mg/L as } CaCO_3$$

452. 24 mg ÷ 12.16 = 1.97 milliequivalents

453. $A$ = 8 mg/L × (74/44) = 13 mg/L
$B$ = 118 mg/L × (74/100) = 87 mg/L
$C = 0$
$D$ = 12 mg/L × (74/24.3) = 37 mg/L

$$\text{Hydrated lime dosage} = \frac{(13 \text{ mg/L} + 87 \text{ mg/L} + 0 + 37 \text{ mg/L}) \times 1.15}{0.90}$$

$$= \frac{137 \text{ mg/L} \times 1.15}{0.90} = 158 \text{ mg/L as } Ca(OH)_2$$

454. Calcium hardness:

$$\frac{x \text{ mg/L}}{50.045} = \frac{31 \text{ mg/L}}{20.04}$$

$$x \text{ mg/L} = \frac{50.045 \times 31 \text{ mg/L}}{20.04}$$

$$x = 77 \text{ mg/L as } CaCO_3$$

Magnesium hardness:

$$\frac{x \text{ mg/L}}{50.045} = \frac{11 \text{ mg/L}}{12.16}$$

$$x \text{ mg/L} = \frac{50.045 \times 11 \text{ mg/L}}{12.16}$$

$$x = 45 \text{ mg/L as } CaCO_3$$

Total hardness = 77 mg/L + 45 mg/L = 122 mg/L as $CaCO_3$

455. Carbonate hardness = 101 mg/L as $CaCO_3$
456. $A$ = 5 mg/L × (56/44) = 6 mg/L
 $B$ = 156 mg/L × (56/100) = 87 mg/L
 $C$ = 0 mg/L
 $D$ = 11 mg/L × (56/24.3) = 25 mg/L

$$\text{Quick lime dosage} = \frac{(6 \text{ mg/L} + 87 \text{ mg/L} + 0 + 25 \text{ mg/L}) \times 1.15}{0.90}$$

$$= \frac{118 \text{ mg/L} \times 1.15}{0.90} = 151 \text{ mg/L as CaO}$$

# Index

# Index

## G



## H



## I



## J



## K



## L



## M



## N



## O

## T



## U



## V



## W



## Y



## Z